Hanno Schmidt-Gothan
Karsten Gessner
Hauke Lübben

The House of Value Creation

Edition Accenture

Edited by Thomas Herbst, Managing Partner Accenture

The series is tailored to meet the information needs of top executives in high-tech, telecommunications and media enterprises. It provides excellent strategy knowledge with target orientation and specialized capabilities. The possible solutions also contain advice on how to implement the required technology and processes in organisations.

Including strategic functions such as shareholder value creation, organisation & reporting, and customer relationship management, the books provide you with top management knowledge and analyse business potential and business models in areas such as media markets, mobile commerce and application service provision.

Recent publications within Edition Accenture:

Interactive Broadband Media
by Nikolaus Mohr und Gerhard P. Thomas

The House of Value Creation
by Hanno Schmidt-Gothan, Karsten Gessner and Hauke Lübben

For more information on coming titles visit our homepage
www.vieweg-it.de

Vieweg

Hanno Schmidt-Gothan
Karsten Gessner
Hauke Lübben

The House of Value Creation

How to Increase Company Value Systematically

vieweg

Die Deutsche Bibliothek - CIP-Cataloguing-in-Publication-Data
A catalogue record for this publication is available from
Die Deutschen Bibliothek.

1st edition August 2002

Vieweg is a company in the specialist publishing group BertelsmannSpringer.
www.vieweg.de

Cover design: Ulrike Weigel, www.CorporateDesignGroup.de
Printing and binding: Lengericher Handelsdruckerei, Lengerich
Printed on acid-free paper
Printed in Germany

ISBN 3-528-05798-X

Acknowledgements

Aside from our individual experiences with numerous projects, the support we received from our colleagues at Accenture significantly contributed to the success of this book. Our thanks go out to Thorsten Brueggemann, Dr. Claus von Campenhausen, Diana Okoye, Arno Pfersdorf as well as Hans-Peter Remark for their valuable input. In addition our free-lance consultant Jens Berwig contributed his advice to this endeavor.

Special thanks are due to our colleagues and editors Dirk von Fabeck, Stefan Hannusch und Alexander Thau. They introduced significant and substantive modifications, proposed additions and corrections, and incorporated them into the manuscript. Without their help this project would not have been possible.

We cordially thank our editor Dr. Monika Luetke-Entrup whose comments significantly contributed to the better understanding and readability of the book.

We also commend Dr. Reinald Klockenbusch, Nadine Vogler-Boecker as well as Walburga Himmel of Vieweg Publishers for their editorial and technical support.

Munich, March 2002

Hanno Schmidt-Gothan

Karsten Gessner

Hauke Luebben

Direct your inquiries to:

Accenture GmbH
Maximilianstr. 35
D-80539 Munich
Germany

Foreword

They have fallen asleep again. What a pity. For a brief period of time, Europe had awoken from its daydream. Entire legions of small investors had become interested in companies and their stocks. A huge wave of business formations and IPOs had followed. At long last, the spirit of Silicon Valley seemed to have arrived at Europe's markets. But the initial wave of broad public interest in stock investing also illustrates that the emerging stock culture and subjects like shareholder value are still in their infancy.

International comparisons show that European, specifically German companies still have a lot of catching up to do. The concept of shareholder value is even used in some cases to connote shame, in order to defend the status quo and to excuse the lack of focus on value creation – reflecting the motto: "There are surely other values aside from stock value" or "We will not bow to pressure from capital markets". However, as the recent take-over of Mannesmann by the British communications company Vodafone shows, the significance of consistent value management clearly cannot be overestimated. The soft stock market environment during the most recent eighteen months reemphasizes the importance of differentiation through systematic value creation approaches.

The stock market needs giants: real-life executive honchos. And also new concepts and performance indicators, which help identify the "hidden champions" among companies. As long as small investors are on their own and have to rely on the traditional information blend of speedy Internet portals, long-winded business reports and emotion-laden analyses of the business press, the stock market and particularly their higher-risk growth segments will remain dormant for a long time to come.

The publication of the House of Value Creation comes at the best time to break down this general lethargy, to rejuvenate the discussion around shareholder value and to cultivate the investment culture in major markets.

The House of Value Creation developed by Accenture is an application-oriented model for integrated management of value creation potential. Based on an easily understandable overall concept, company upper management and observers can address pivotal strategic issues: What vision accompanies a corporation on its quest in the market? How does it control growth and profitability? How solid is the corporate foundation? Are strategy and operations aligned towards maximizing value creation? Metrics and performance targets for business architecture and organization need to ensure that competitive advantages are consistently utilized and that company philosophy is characterized by customer-orientation and closeness to markets to the fullest extent.

So the operative expression is value management. However, this should in no way be confused with the traditional key performance indicator (KPI) analysis. Considering the fast innovation rate and high market volatility for instance in the technology sector, relying exclusively on historical financial data represents an unreliable aid in planning for the future. Even more so since decisive assets of modern companies are not even considered in traditional financial reports – be it intellectual capital, innovation potential, customer orientation, the ability to effectively integrate acquisitions, quality of cooperation with suppliers, learning culture, etc. However, the integrated evaluation of business logic and fundamental values captures precisely these levers toward systematic value creation.

The House of Value Creation supplies the readers with an instrument that will allow them to approach companies via a completely new venue. At last, ideas and concepts are presented, which give the analyst as well as the interested investor a forward-looking basis for evaluation. But they also support company executives in obtaining an overall picture of their organization from a holistic viewpoint and in implementing an integrated approach towards further value creation.

It is precisely this quality, which characterizes successful companies: the overall picture is convincing in all its details – starting at the value-creation-oriented vision, through management of sustainable growth, to inclusion of all the sources of innovation. The House of Value Creation is a simple and elegant model that explains interrelations and corporate significance of shareholder value, while relating it to corporate management in a comprehensible and hands-on fashion. The management of the main levers of value creation is explained in a comprehensive, compelling, and – in my opinion – practice-oriented manner.

Tenovis, too, – a company born in the Old Economy, on its quest towards an innovative, value-creation-oriented future – has successfully institutionalized and implemented into daily business practice numerous concepts and ideas generated in the course of its cooperation with Accenture. We are excited to see that these innovative concepts are presented here in a compact and application-oriented form. This book is a genuine enrichment of the discussion on shareholder value and it should become mandatory reading for company executives, analysts, institutional investors, consultants, students of economics and involved individual investors.

Frankfurt, June 2002

Peter B. Záboji, CEO Tenovis

Peter B. Záboji (57) studied Physics and Economics in Germany, Belgium and France. Initially he worked for IBM. He has held several executive positions at Siemens USA and Europe. In the early 90s, Záboji needed only nine months to weld six British competitors into GPT Communications, the largest privately owned telecommunications company in the country. In the mid-nineties, as managing director of o.tel.o, he was responsible for the successful market introduction of a new brand. Simultaneously Záboji was the chairman of the board of germany.net and director of E-Plus. Starting in 1999 he accompanied and supported young entrepreneurs as their "Business Angel" in the company-founding phase.

Since April, 2000 Peter B. Záboji has been at the helm of Tenovis. Within the shortest time, he transformed Tenovis from a German manufacturer of telephone equipment into a European client-centric provider of solutions and services for business communications.

Table of Contents

Introduction...1

1

Value Creation: Coincidence or Strategy?..5

1.1   Key Determinants of Value Creation – Endogenous or Exogenous Factors?.5

1.2   The Neuer Markt – "Salami Crash" and Few Winners..................................7

1.3   Value Creation Patterns in the High-Tech Industry....................................10

1.4   Cisco – Industry leader through growth and profitability..........................15

2

The House of Value Creation: An Integrated Model for Sustainable

Value Creation...19

2.1   Company Vision and Aspirations...20

2.2   Value Creation Levers..21

2.3   Metrics and Incentive System...22

2.4   Organizational Foundation..23

3

Company Vision & Aspirations: The Roof of the House............................27

3.1   Clear Visions are a Prerequisite for Value Creation....................................27

3.2   Aligning Heterogeneous Aspirations Towards the Goal of Value
        Creation — A Pragmatic Guide..32

   3.2.1   Individual Interviews with Members of Senior Management..............33

   3.2.2   Qualitative and Quantitative Evaluation...34

   3.2.3   Plausibility Analysis of the Results...39

   3.2.4   Feedback and Goal Specification Workshops......................................42

   3.2.5   The Result: GreatValue, Inc. on the Road to Doubling its Valuation...44

4 Value Creation Levers: The Supporting Pillars.............................47

4.1 Managing for Profitability or Growth?...47

    4.1.1 Significance of the Levers as a Function of the Company Situation.....47

    4.1.2 Value-Benchmarking Against the Best-in-Class – Not Just Profitability Counts...50

    4.1.3 Integration of Both Pillars into a Successful Value Creation Program..53

4.2 Increasing Profitability Through Cost Optimization.........................55

    4.2.1 Determination of the Optimization Potential.........................57

        4.2.1.1 Comparison With the Best-in-Class..............................57

        4.2.1.2 Functional Benchmarking...63

        4.2.1.3 Detailed Drill-down Analyses....................................69

    4.2.2 Design of Program Components...70

    4.2.3 Development and Adoption of Detailed Measures.......................72

    4.2.4 Implementation and Monitoring.......................................74

4.3 Securing the Future Through Growth..77

    4.3.1 Managing Growth Horizons – Balance Ensures Sustainability.............77

        4.3.1.1 Growth Option Pipeline and Horizons............................77

        4.3.1.2 Growth Options..82

    4.3.2 Venues to Growth..85

        4.3.2.1 Organic Growth..86

        4.3.2.2 Growth Through Venturing..88

        4.3.2.3 External Growth...90

5 Metrics and Incentive Systems: The Supporting Floor.......................97

5.1 Operationalization of Corporate Objectives Through Concrete Metrics......97

5.2 Principles of the Market Economy as a Basis for the Evaluation of Corporate Entities..99

    5.2.1 Profit Centers..100

5.2.2   Transfer Prices 100

5.3   Metrics as Performance Indicators and Benchmarking Standards.............. 103

5.3.1   Metrics for Value-Based Management 103

5.3.2   Individual Performance Indicators for all Employees 105

5.4   Incentives to Reach Target Metrics... 106

6

The Organizational Foundation: The Basis of the House of Value

Creation 109

6.1   Portfolio Structure 109

6.1.1   Strategic Portfolio Optimization – Increasing Transparency.............. 110

6.1.2   The Generation Conflict – The Daughter Becomes Independent 112

6.2   Business Architecture... 115

6.2.1   Optimization Levels – Organization, Processes and Systems 115

6.2.2   Optimization Approaches 116

6.2.2.1   Process Optimization.. 119

6.2.2.2   Creation of Shared Services.. 121

6.2.2.3   Introduction of ePlanning / eReporting............................ 123

6.2.2.4   Project Portfolio Optimization..................................... 126

6.2.2.5   Efficient Resource Planning in Research and Development 128

6.2.2.6   eProcurement System Implementation 134

6.2.2.7   Design-to-Cost 136

6.2.2.8   Value Chain Optimization 138

6.2.2.9   Efficient Supply Chain Management (SCM)........................ 141

6.2.2.10   Systematic Customer Relationship Management...................... 143

6.3   Corporate Finance... 146

6.3.1   Cost of Capital Reduction... 146

6.3.1.1   Capital Structure Balance.. 146

6.3.1.2   Central Capital Management 148

6.3.2   Funding Growth 149

6.3.2.1   Funding Growth with Internal Capital.............................. 149

6.3.2.2   Funding Growth with External Capital 151

6.3.2.3   Funding External Growth 152

6.4   Human Capital 156

6.4.1   Personnel – More Than Just a Cost Factor 156

6.4.2   Human Resource Management Phases 157

6.4.2.1   Recruiting Employees 157

6.4.2.2   Staffing Positions and Employee Reassignment 160

6.4.2.3   Measuring and Rewarding Performance 160

6.4.2.4   Promotion of Development and Continued Education 162

6.4.2.5   Employee Retention 163

6.4.2.6   Knowledge Management 165

6.4.2.7   Termination Procedures 166

6.4.3   Analysis and Elimination of Performance Deficits 167

6.5   Investor Communication 169

6.5.1   Investor Awareness – The Key to Valuation 169

6.5.2   Investor Communications Audiences and Instruments 171

6.5.2.1   Financial Analysts as Opinion Leaders 171

6.5.2.2   Sustainable Value Creation Through a Credible Equity Story 173

6.5.2.3   Investor Awareness and Accounting Standards 175

6.5.3   Expectation Management is at the Heart of Successful
Communications 176

6.5.3.1   Exceeding Expectations Yields Value Creation 176

6.5.3.2   Managing the Expectation Treadmill 177

7   The Complete House: Holistic Value Creation in Practice 181

7.1   Program Management – A Holistic Implementation Approach 184

7.2   Case Study – The Example of GreatValue, Inc. 188

7.2.1   Phases 1 and 2 – Analysis of Value Creation Potentials 188

7.2.2   Phases 3 and 4 – Implementation of the Value Creation Program 190

8   Outlook – Strengthening the Corporate Future 193

Appendix 195

    A: Value Creation – Relative Valuation Map 195

    B: Using the DCF Method to Derive Company Values 197

Table of Figures 203

Glossary 209

Index 217

Bibliography 221

Introduction

"The only moral responsibility of a company is to create value". This is how the American economist and Nobel laureate Milton Friedman once summarized his views on the business world. Indeed, though there are other legitimate responsibilities for modern management, value creation is the most significant objective by far. It not only satisfies investors, but primarily ensures growth, independence and thus continued corporate existence. Value creation is possible across industries and does not occur accidentally. And, of course, "shareholder value" systematically creates "stakeholder value".

Even though many corporations define value creation as their central objective, only a select few actually manage to align the whole organization towards this goal. An integrated approach that strives towards anchoring value creation as the strategic and operative maxim throughout all corporate sectors and finally also implementing it in real life is rare. Particularly in dynamic industries metrics like *Economic Value Added (EVA)* often fall short of achieving a holistic alignment of the corporation towards value creation.

These approaches, which are either too abstract or too narrowly focused and do not complement each other across all corporate areas often lack orientation towards growth and towards the future.

Project experience demonstrates that a *tactical gap* frequently opens between global strategic objectives and concrete measures of implementation. The operationalization of value creation strategies – for instance in regard to implications on business plans, product and manufacturing strategies, or the alignment with controlling instruments – is frequently neglected. Also the alignment of different elements of the business model (aspirations, target metrics, controlling mechanisms, processes and systems) in many value creation programs is instituted only insufficiently or not at all. This in turn causes value creation potentials not to be exploited optimally.

Only a holistic approach assures that value creation potentials are utilized to the fullest extent and are realized in a sustainable fashion.

The first step to value creation is the value benchmarking of a given corporation versus the industry leader. If a value gap is uncovered, a precise diagnosis of its causes needs to be carried out. Only then can value creation programs and their main directions of thrust be formulated logically and synergistically. Only comprehensive programs without wasteful redundancies lead to maximal value creation.

Certain success patterns are repeatedly observed at value leaders across a variety of industries. Accenture has systematized these observations and has developed a comprehensive management instrument on this basis – the House of Value Creation.

The House of Value Creation represents an integrated and holistic approach for the diagnosis of value creation opportunities and the corresponding derivation of value creation programs. This approach aids in the identification of value creation potentials and the obstacles threatening their realization. It allows the definition of operative strategies for bridging gaps between corporate strategy and planned objectives, and it helps in designing integrated implementation programs.

The orientation along three dimensions illustrates the comprehensive nature of the House of Value Creation:

1. The corporate target value (most frequently externally imposed, e.g. by the stock market),
2. the objectives of corporate management and,
3. the corresponding prerequisites and levers required for implementation, such as business strategy, metrics and incentive systems, business architecture as well as individual value creation initiatives.

The House of Value Creation model is not only suitable as a management tool but also as a diagnostic instrument for analysts and investors. In order to determine a company's true value creation potential a specific evaluation of business logic and fundamental values is necessary in addition to mere financial analysis. The most recent stock market crash has clearly shown that financial data, ratios, and other quantitative indicators alone are not reliable guides for the prediction of future developments. Only the inclusion of fundamental business logic produces a meaningful overall picture.

Numerous reputable corporations have been applying the House of Value Creation for successful value creation. Based on the continuous collection and evaluation of corporate data Accenture has refined the model methodically and supplemented it with practical insights.

This book presents the House of Value Creation as an instrument for comprehensive company analysis and for holistic alignment towards value creation. Numerous examples and cross references demonstrate a pragmatic venue for value maximization to managers, analysts, consultants and also to students of economics.

The book is divided into eight chapters:

Chapter 1 analyzes – representative of the entire spectrum of industries and markets – companies in the high-tech sector on a global scale as well as the Neuer Markt (Germany's stock exchange for high growth and technology companies, Europe's most important growth market). It identifies and systematizes success factors among companies that have achieved above-average value creation during the observation period. These success patterns provide the basis for the elements, which make up the House of Value Creation.

Chapter 2 introduces the concept and structure of the House of Value Creation. The individual components of the House are then analyzed in greater detail in the following chapters.

Chapter 3 analyzes the significance of the company vision for corporate value creation. In addition, it will demonstrate how heterogeneous personal aspirations of the members of upper management can be aligned with this vision.

Chapter 4 focuses on the two most important levers of value creation: profitability and growth. It details different methods that can be employed to first quantify profitability and growth gaps and then to close them by using a suitable combination of the value creation levers.

Chapter 5 explains how a value creation program is implemented on an individual as well as on an institutional level with respect to appropriate controlling mechanisms, metrics, and incentive systems. In this manner it illustrates a venue via which corporate units and individual employees can optimally support the value creation process.

Chapter 6 outlines concrete value creation measures in the areas of portfolio structure, business architecture (organization, processes, and systems), financing, human resources management, and investor communications.

Chapter 7 demonstrates the integrated application of the House of Value Creation in addition to the institutionalization of value creation processes based on a real-life example. This chapter also shows the necessity of an integrated program management.

Chapter 8 explains how, particularly in times of unfavorable economic prognoses, systematic value creation serves as a differentiating factor.

The reader will find continuously updated examples and articles for discussion as well as an opportunity for a constructive dialogue with the authors at:

www.house-of-value-creation.com.

Your comments are very welcome.

1   Value Creation: Coincidence or Strategy?

The boom years on the stock exchange of the late nineties have come to a sudden halt recently. The drastic loss in value of almost every internationally listed stock provokes one critical question to investors and managers alike:

Was this loss in value predetermined by external factors, or could companies nonetheless create value in this challenging environment through intelligent management efforts focused on systematic value creation?

1.1   Key Determinants of Value Creation – Endogenous or Exogenous Factors?

Many companies listed at the Neuer Markt (New Market)– Germany's stock exchange for high growth and technology companies – one-sidedly emphasized fast growth in their early stage of development instead of pursuing profitable value creation. This worked well during the boom, until March 2000. At that point many investors suddenly demanded real value creation and thus caused the collapse of many of these growth stocks. On closer scrutiny, problems also surfaced at well-established companies in other industries. Nevertheless, some companies like Cisco Systems, General Electric, Intel Corporation and Sun Microsystems succeeded in significantly increasing their market valuations even during this negative stock market cycle.

Structure of the Analysis

The value of a company is not only determined by the performance of its management. Several exogenous factors, particularly the relevance of the company's industry affiliation are frequently mentioned in literature. An extensive empirical analysis of the value created on the capital markets conducted by Accenture clearly shows whether above-average value creation is limited to individual industries and market segments, and whether it is due to "coincidence" or strategy. This analysis has been conducted along two dimensions, as illustrated in figure 1-1.

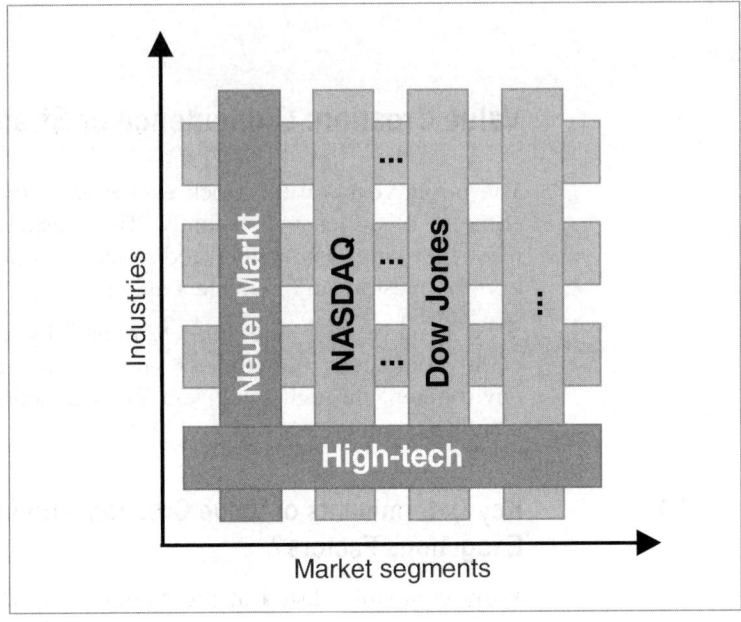

Figure 1-1:   Dimensions of the capital market analysis

The literature generally assumes that there are stark contrasts between the best-in class and their industry peers in dynamic markets. The high growth and differentiation potential of these industries are often listed as the main reasons. Accenture empirically challenged these assumptions. Several studies analyzed whether dependencies exist between value creation and industry affiliation for start-up companies in "young", i.e., emerging, industry sectors. Specifically, the development of market values in different industries within the Neuer Markt was examined. The results of these studies are presented in chapter 1.2.

Analysis of the High-tech Industry Across all Market Segments

Accenture also verified whether similar dependencies exist for well-established companies in different capital markets. The high-tech industry was particularly suited for this analysis. Different segments display continuously high but very distinct dynamics. Although the analysis started with the total market, the focus was on the established companies. To cope with the broad spectrum of high-tech companies, the market was divided into the following representative segments: semiconductors, computers, network suppliers and industrial electronics. These segments were then analyzed in the search for value creation patterns. The results are described in chapter 1.3.

This coverage of both dimensions – capital market segment and industry affiliation – permits the derivation of universally valid statements that also hold true for the intersections that have not been analyzed.

1.2 The Neuer Markt – "Salami Crash" and Few Winners

Since March 10, 1997 the Neuer Markt exists as a trading segment of the Deutsche Boerse (the German Stock Exchange) with a high opportunity and risk profile. The idea is to facilitate access to the capital market for young and innovative growth companies. The high degree of transparency and the comparability of companies listed at the Neuer Markt due to standardized rules are particularly suited to attract sufficient numbers of investors.

Development of the Neuer Markt

All stocks of the Neuer Markt are quoted on the Nemax All Shares index. The vastly growing number of participants and their steadily increasing market values prompted the German stock exchange to introduce the Nemax 50 index in July 1999. This index represents the 50 largest companies of the Neuer Markt according to their market capitalization and trading volume. It was particularly intended to increase this stock market segment's attractiveness to institutional investors. To enable a comparison with the Nemax All Shares index, the Nemax 50 was adjusted to account for the period, which had elapsed since July 1997, and normalized to a base value of 1,000 points as of December 1997. Industry indices were likewise introduced in May 2000. Investors can now compare the development of companies within an industry as well as the development of the Neuer Markt industries.

Even in the early weeks and months of 2000 companies listed at the Neuer Markt forecast steeply growing revenues and quickly reaching their break-even point. The majority of the companies could not live up to these exaggerated expectations. Dramatic declines in corporate revenues and profits coupled with news scandals led to a sharp drop in stock prices. In the aftermath, shaken investor confidence caused the continuous erosion of stock prices on the Neuer Markt.

The extent of the damage is illustrated in figure 1-2. In March 2000 the Nemax All Shares had reached its peak of almost 9,000 points. During the following 15 months, through July 2001, the index slumped to below 1,500 points. The Nemax 50 moved in step with its parent index. In September 2001 the quotes of both indices dropped well below their base values, to less than 800 points.

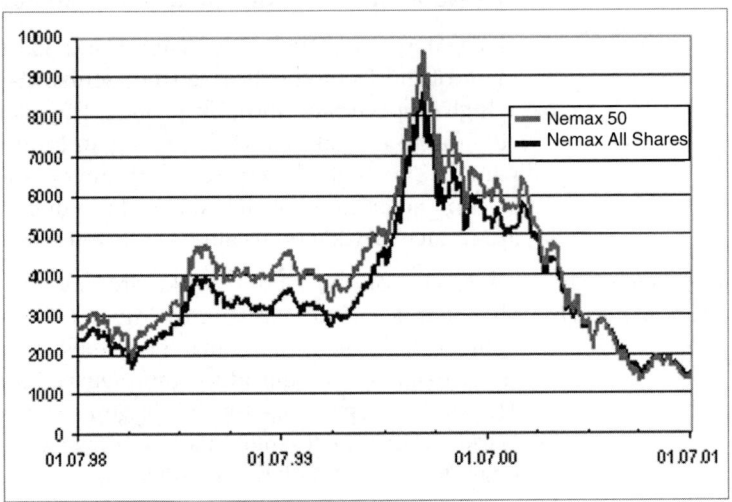

Figure 1-2:   Development of the Nemax All Shares and Nemax 50 indices

Despite this erosion on the Neuer Markt, some industries managed to achieve overall positive returns since July 1998. Particularly the Biotech segment, with an average annual return of 37 percent, outperformed all other sectors. But these positive developments of individual industry segments were not carried by all of the companies equally: There were winners in all segments, which were able to detach themselves from the negative overall industry trend. (figure 1-3).

Nemax Index	Return on Capital				Total
	< -60%	-60% to -11%	-10% to +10%	> +10%[1]	
Biotechnology	3	9	2a	7	21
Financial Services	0	4	0	1	5
Industry & Industrial Services	4	7	2	6	19
Internet	41	22	1	2	66
IT Services	14	15	5	3	37
Media & Entertainment	16	20	4	1	41
Medical Technology & Health Care	3	6	3	1	13
Software	23	19	3	1	46
Technology	13	24	10	26	73
Telecommunications	10	9	0	1	20
Total	127	135	30	49	341[2]

Source: Accenture Research, Datastream

[1] Highlighted column: leader segment
(Assumption: 10 percent minimum annual return on capital)

[2] 64 of 341 companies are listed less than twelve months on the Neuer Markt. Their return was extrapolated to a one-year period.

Figure 1-3:   Number of winners and losers by industry

Assuming average capital costs of ten percent, only 49 of 341 companies traded on the Neuer Markt have created value. Only one winner could be identified in each of the two industries that performed most poorly during the period under consideration - software and telecommunications. These out-performers were FJA (software) with a 26 percent average annual return and Funkwerk (telecommunications) with 91 percent. It is remarkable that both companies had their IPO while the Neuer Markt was soaring and still managed to create value in the face of the succeeding developments. Other examples of successful companies are Broadvision (internet), Direkt Anlage Bank (financial services), Computerlinks (IT services), Deag Deutsche Entertainment (media & entertainment) and Eckert & Ziegler (medical technology & health services). They also achieved returns of more than 10 percent despite the free-falling trends that characterized their overall market segments.

1.3 Value Creation Patterns in the High-Tech Industry

The four segments of the high-tech industry that were analyzed in the second phase – semiconductors, computers, network suppliers and industrial electronics – represented worldwide revenues of US$900 billion in 2000. The indicator *Exceeding Economic Expectations* (E³; Box 1) is particularly suited for evaluating their value enhancement performance.

Box 1: Total return to shareholders beyond investor expectations – Exceeding Economic Expectations

Investors invest in a company if the expected total return is high enough to compensate for the company-specific risk. According to the *Capital Asset Pricing Model (CAPM)*, the expected investor return – equal to the company's cost of capital – equals the return of a risk free investment plus the company-specific risk premium. Similar to the *Economic Value Added (EVA)* – the measure of the positive business value contribution – Exceeding economic expectations (E³) measures the excess return of a stock over and above investor expectations (figure 1-4).

To calculate E³ the cost of equity capital (COE) and the residual market risk (RMR) are subtracted from the total return to shareholders (TRS):

E³ = TRS – COE – RMR. The total return to shareholders equals the stock price change (in percent), adjusted for cash flows to investors. Examples of such cash flows are dividends or stock bonuses. The residual market risk cannot be influenced by management and comprises changes of interest- and exchange rates, or speculative stock movements. E³

is above all influenced by consistent management performance and can therefore be used as an evaluation tool.

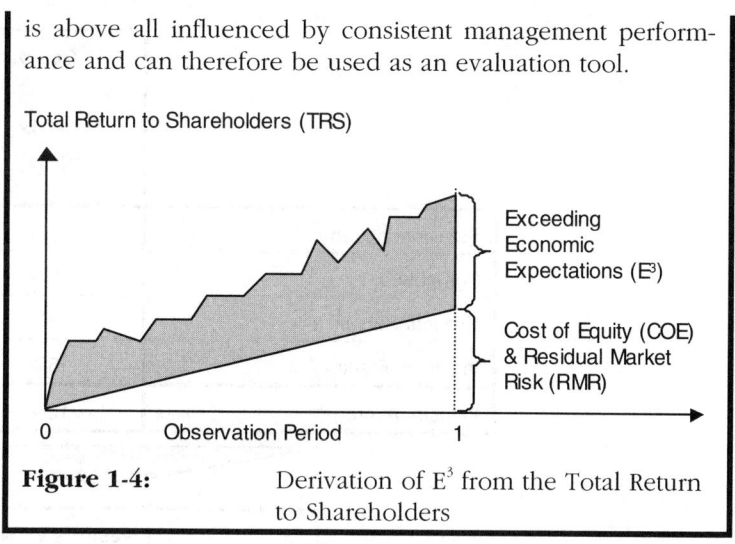

Total Return to Shareholders (TRS)

Exceeding Economic Expectations (E^3)

Cost of Equity (COE) & Residual Market Risk (RMR)

0   Observation Period   1

Figure 1-4:   Derivation of E^3 from the Total Return
                         to Shareholders

Figure 1-5 compares different indicators for value creation across
the four analyzed industry segments for the period from June
1998 to June 2001. These average values are based upon performance indicators of selected companies, weighted according
to their respective share of segment revenue. All four industries
show discernibly different value creation potentials. Note that industrial electronics (with an E^3 of more than five percent) is the
only industry segment that outperformed expectations. This does
not mean that no individual companies in the other segments
were able to create value as measured by E^3, since these are average values for the entire segments. It does indicate, however,
that these segments, on average, could not fulfill expectations.

A representative selection of companies from every segment was
formed and their performance was analyzed in order to establish
whether individual companies could escape the general trend of
their industry segments. Selection criteria were market capitalization, revenue growth and the dynamics of their business development between June 30, 1998 and June 30, 2001.

	Company Parameters	
	Return on Equity	Revenue Growth
Semiconductors	19%	7%
Computers	16%	9%
Network Suppliers	12%	11%
Industrial Electronics	15%	11%
Industry Average	16%	10%

	Investor's View		
	Total return to shareholders	Capital cost	Exceeding Economic Expectations (E³)
Semi-conductors	7%	11%	-4%
Computers	8%	10%	-2%
Network Suppliers	-2%	11%	-13%
Industrial Electronics	15%	10%	5%
Industry Avg.	5%	11%	-6%

Figure 1-5:   Performance indicators of the four high-tech industry segments analyzed

The market capitalization of a company is an extremely relevant figure. It measures the influence and the strategic reach of the company in a specific market, i.e., the potential to influence the industry's structure by acquiring other companies.

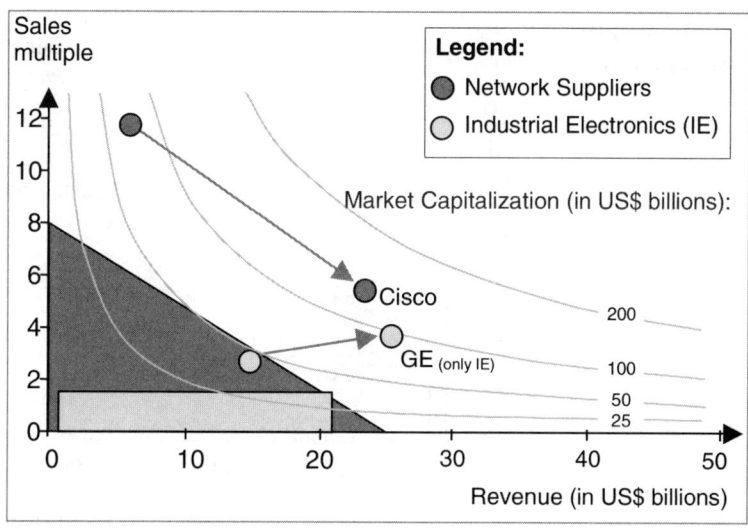

Figure 1-6: Relative valuation map of the Network Suppliers and Industrial Electronics segments in June 2001

The relative valuation map in figure 1-6 and figure 1-7 shows the capital market positions of the selected companies and industries. To maintain comparability, the segment-specific revenues of conglomerates were used in addition to the market capitalizations weighted by the revenue categories (sum of the parts model). We can observe that some companies (value leaders) created substantially more value than competitors within their own industry and sometimes even more than companies from industries with inherently higher value creation potential.

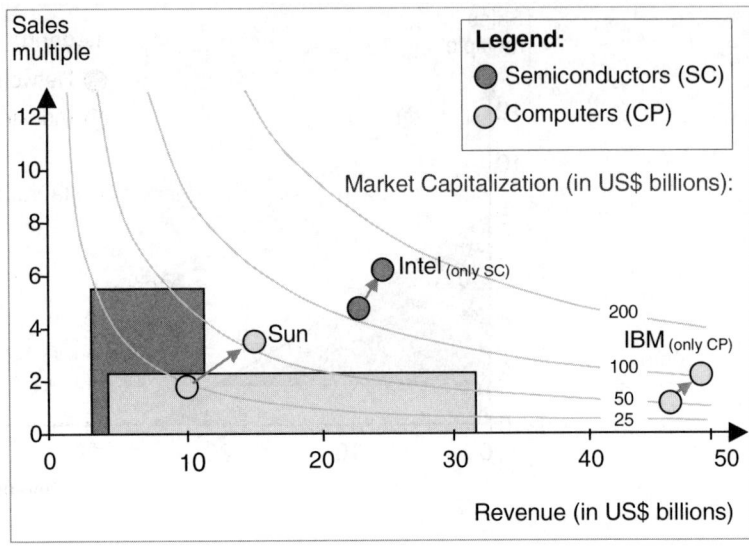

Figure 1-7:   Relative valuation map of the Semiconductors and Computers Segments in June 2001

The relative valuation map – explained in appendix A – does not directly permit the derivation of the value the companies have created from June 1998 to June 2001 because it only represents a snapshot view. Therefore figure 1-6 and figure 1-7 also include the migration paths which the selected companies have traced during the past three years and along which they have shifted their strategic positions.

It shows that value leadership in general forms a solid basis for value creation but that other companies can also achieve higher valuations by rigorously focusing on the relevant value creation levers.

The analyses reveal that some companies have been able to distance themselves from both the industry- as well as the market trends - and thus create significantly more value. It follows that value creation is not necessarily dependent only on the industry- or market segment. The Neuer Markt represents a cross-industry capital market. Companies of the high-tech industry are represented on various capital markets. Thus the conclusions and recommendations of the two analyses can ultimately be applied to all companies in all capital markets.

1.4 Cisco – Industry leader through growth and profitability

Based on the example of one of the high-tech industry's best-in-class companies, Cisco Systems, we will subsequently illustrate company-specific value creation achievements.

Cisco Systems was founded in 1984 by a group of computer experts at Stanford University. In 1986 Cisco introduced the first router that could combine heterogeneous networks. After its IPO in 1990 Cisco evolved into the world leader in network systems and infrastructure within a few short years. Cisco's product range covers hardware and software as well as services. Today, Cisco is one of the largest companies in the world – in terms of market capitalization – and controls more than 75 percent of the world market for routers and switches. Cisco currently employs approximately 40,000 people and revenues were US$22.2 billion in fiscal year 2001.

Vision

Cisco's business is permeated by the vision that the Internet is changing the way we live, work, learn, and spend our leisure time. Cisco wants to actively shape the future of this medium by developing new opportunities for its customers, employees, investors, and partners by going beyond its own business. The company intends to remain number one in the network infrastructure business. Furthermore Cisco strives to attain a target market share of 40 percent in each market it enters and a minimum level of 25 percent. To attain its ambitious goals, Cisco builds on continuous improvement of its profitability and sustainable, high growth rates. Due to the "constant paranoia" propagated by Cisco's CEO John Chambers, the company does not rest on the laurels of its achievements but aggressively strives for continual growth.

Profitability

Cisco has always been insistent on establishing excellent structures and processes across all areas of its business and on ensuring the greatest possible transparency. Therefore the company does not shy away from restructuring the organization, rationalizing its product range or realigning its resources. In particular Cisco extensively relies on the use of the Internet. Not only is information made available via the company's intranet. The Internet is also employed to conduct transactions. Web-based order processing pushed down the rate of erroneous orders from 33 percent to less than 2 percent. Utilization of online help pages reduced expenditures for technical support by US$ 173 million in fiscal year 2000. In terms of resource management, the integration of the supply chain and outsourcing of production are the next targets.

Growth Strategy

Cisco's growth strategy depends on a combination of organic and external growth. The basis for its organic growth is constant improvement of current and development of new products, as well as extending existing businesses and capturing new markets. New products, distributed in new markets, are supposed to generate 30 percent of total growth. The focus is therefore on markets that are anticipated to grow explosively.

Accordingly, Cisco is constantly striving to shorten the development time of its products. In return Cisco is willing to accept that not all the product flaws are eliminated upon market entry. This turning away from perfectionism has obviously been very successful: with a sound new product, potential flaws could be eliminated through close cooperation with the customer. With this strategy Cisco molds the dynamics of its industry segment and is capable of reacting quickly and flexibly to changes in the marketplace.

In addition to organic growth, Cisco acquired more than 70 companies since 1993 to fill gaps in its product line, to expand its distribution channels, or to enter new markets. Cisco demonstrated remarkable skills with the integration of its acquired companies. Usually integration was completed within 100 days. The key talents of the acquired company could be integrated into Cisco's corporate culture and were able to significantly contribute to value creation at Cisco very quickly. Apart from acquisitions, Cisco also targets alliances, strong partnerships and joint ventures to supplement its own product development efforts and to enhance market penetration.

Metrics

An integral part of the company's philosophy is to manage the departments and employees by means of metrics that are aligned with the company's corporate targets. Cisco consequently manages its employees using a bonus system. Each employee is motivated to set visionary, specific and above all measurable goals for himself/herself. The achievement of objectives is not only measured by quantitative factors, e.g. revenue and profit, but also by qualitative ones.

Employees have a high degree of autonomy in the fulfillment of their individual ambitions. They are able to focus on things they can influence. This incentive system is the reason why many employees identify themselves with the company and align their personal goals accordingly. The result is growing customer and employee satisfaction.

Derivation of the House of Value Creation

Cisco has attained a very high market capitalization by focusing on growth and profitability. This in turn enabled further aggressive expansion by supplying the company with a currency for stock-swap based acquisitions. Cisco was able to outpace both the industry and the market trend and steadily increase its market value by consistently aligning its employees towards the company's goals, by driving product innovation, and by lowering costs.

Cisco is just one example of many. Accenture has analyzed a broad spectrum of success stories, evaluated their business strategies, architectures and structures and systemized them into the House of Value Creation. This model describes an approach by means of which companies across all industries can identify the success factors for value creation, realize value creation potentials, and thus increase their market capitalization.

2

The House of Value Creation: An Integrated Model for Sustainable Value Creation

Value Creation Through Holistic Management

The goal of the House of Value Creation is to instill a profound and ongoing understanding of continuous value creation in companies. It provides managers with analysis tools for value-based benchmarking, enabling them to identify valuation gaps to best-in-class competitors and to better understand the causes of these gaps. Furthermore the House of Value Creation illustrates the various elements and steps of a comprehensive value creation program and offers concrete recommendations for its practical implementation.

The holistic approach advocated by the House of Value Creation is structured similarly to the *Total Quality Management* program that was developed in the eighties. Managers of the automotive industry above all succeeded in engraining quality as a company goal in the employees' mindset. The intention behind The House of Value Creation is to initiate an analogous continuous process in which the environment in companies is consistently updated and improved to become even more conducive for value creation.

The model integrates all areas of value creation. Specific methods for maximizing value are assigned to each element of the house. Depending on the individual situation of a company, specific recommendations for action are generated for the implementation of a comprehensive value creation program.

Levels of the House of Value Creation

The House of Value Creation consists of four levels (figure 2-1):
1. Company vision and aspirations
2. Value creation levers
3. Metrics and incentive systems
4. Organizational foundation

Concrete value creation potentials can be discovered on each of these levels by means of key questions. This way a company can analyze in detail which barriers to value creation exist and what measures are necessary to overcome them.

19

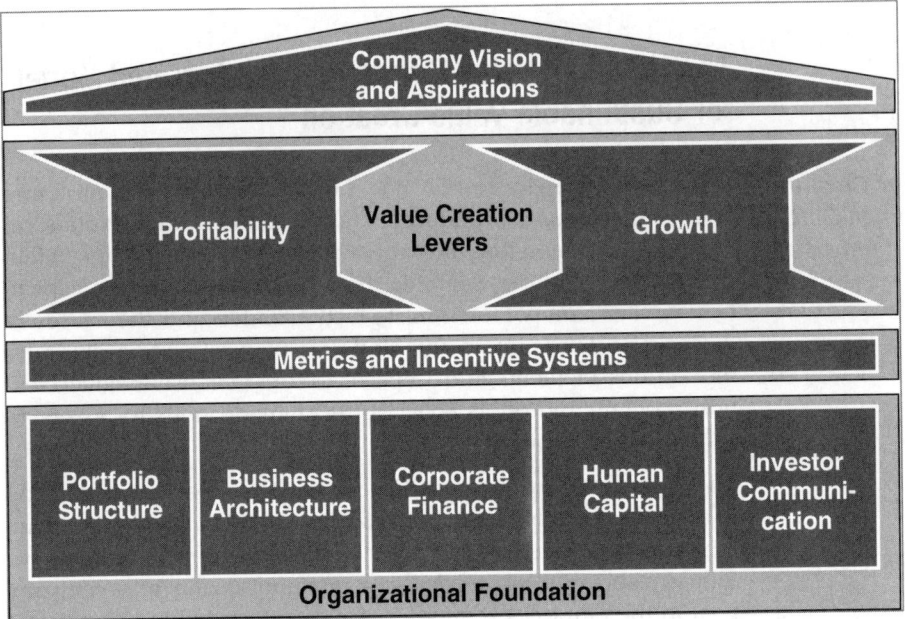

Figure 2-1: Blueprint of the House of Value Creation

2.1   Company Vision and Aspirations

The vision describes the overall goals of the company. It influences the core business as well as the strategic orientation. A company vision should have longer-term validity and should name the increase of the company value as a central goal either implicitly or explicitly. Furthermore, the personal (and sometimes very heterogeneous) aspirations of management need to be harmonized and aligned towards the realization of the vision.

The following key questions demonstrate the scope of the House of Value Creation's top layer:

2.2   Value Creation Levers

Profitability and growth are the levers, which determine a company's capital market valuation. This valuation especially includes investor expectations about the future development of these levers. The relative importance of both levers strongly depends on the maturity of the industry and the development stage of the individual company. As a rule of thumb, profitability grows in importance the more mature an industry is. In dynamic markets – in the early stages of industries – growth is more important. But growth at any cost, on the other hand, is not the way to build a sound business. Numerous examples from start-up companies and the high tech industry demonstrate this point.

Profitability

The best-in-class create exemplary, cross-functional organizational structures on the basis of which the productivity level can be increased. Rigorous cost management is particularly important in this regard. It is therefore critical to first identify those processes that have the highest cost reduction potential in comparison to competitors. This in turn presupposes a thorough understanding of the internal cost structure and a systematic benchmarking with the best-in-class. Afterwards measures have to be developed with which the identified areas for improvement can be optimized. Examples of such measures are design-to-cost or outsourcing initiatives. These programs affect the entire business architecture – the company's organization, its processes and the deployed systems. A high level of program detail leads to increased reliability in implementation. A project monitoring structure verifies the fulfillment of program targets.

Growth

Instead of solely reacting to market changes the best-in-class proactively realign their company's strategy, for instance, by diversifying into new markets. Whereas Nokia primarily manufactured paper and rubber products for the Finnish market only 20 years ago, the company now is a leading provider of mobile telecommunications equipment.

The best-in-class constantly re-evaluate the range of their products and business activities and systematically develop their future product portfolio. Innovation capabilities, short development cycles and product introduction strategies are of special strategic importance. Moreover these companies are able to acquire and integrate companies within the shortest possible time frames. Marketable products, commercializable know-how and the acquisition of experts are central to these takeovers – not machines and factories.

A first impression about whether a company optimally employs the value creation levers growth and profitability is gained through answering the following key questions:

Key Questions: Value Creation Levers

➤ How big is the value gap to competitors and what are its causes – growth deficits, profitability gap or both?

➤ Can the company manage costs even in times of rapid growth?

➤ Does the company continuously review and develop its product portfolio or does it rely on few products with a short half-lives?

➤ What mechanisms and what financial potential exist to acquire companies speedily with a minimum of frictional losses?

➤ For start-ups: Is a concrete and realistic break-even point on the horizon or is growth paid for with a downward spiral of losses?

2.3 Metrics and Incentive System

The alignment of business units (institutional management) as well as of personal aspirations (individual management) towards the company's goal of value creation is indispensable. Institutional incentives should steer the employees' individual needs in such a way that the decision makers on every functional level are motivated to define value creation as their personal goal.

Successful companies manage each business unit by its contribution to value creation and tie parts of the executives' salaries to the units' performance. The introduction of a *balanced scorecard* offers an efficient way to evaluate non-quantifiable metrics. Efficient use of incentive systems may be checked according to the following key questions:

Key Questions: Metrics and Incentive Systems

➢ By which metrics is the company managed?

➢ Are the incentive systems efficient and aligned with the metrics?

➢ Are these metrics also defined as individual goals for the employee?

➢ Is an incentive system in place that motivates outstanding performance and also supports non-quantifiable metrics?

2.4 Organizational Foundation

A solid organizational foundation is necessary to optimally employ the value creation levers. This foundation consists of portfolio structure, business architecture, company finance, human capital and effective investor relations.

Portfolio Structure

The degree of centralization, the autonomy of individual business units and the structure of the business portfolio play an important role in the market valuation. Conglomerates must try to increase their transparency in order to reduce or even avoid a conglomerate discount in the capital markets. Otherwise, the market valuation of a few small, aspiring companies might exceed that of a well-established company. Factors like tightly interwoven corporate structures or cross-subsidization are less relevant for young companies than for conglomerates. Still, even in dynamic industries obfuscating redundancies might arise quickly as a result of mergers or acquisitions.

Business Architecture: Organization, Processes and Systems

Organizational structure, processes and systems should always be oriented towards the greatest possible customer orientation and closeness to market.

Though flat hierarchies increase flexibility and sense of responsibility, competencies must be clearly defined. This is true for established companies but also for start-ups. Moreover the positioning within the industry's value chain must be clearly defined and optimized in regard to the core competencies. *customer relationship management (CRM), supply chain management (SCM),* process and product optimization measures as well as outsourcing are exemplary starting points for such an optimization of the business architecture.

Corporate Finance

An optimal financial structure allows for maximum profits at constant capital expenditure and for profitable investments at minimal costs of the capital employed. A reduction of the capital employed frees resources for other initiatives. Companies with sound financial management are thereby able to generate opportunities for further growth.

Human Capital

All phases of a work relationship – from recruiting to retirement – should be organized systematically. The recruiting and retention of highly qualified employees and the acquisition of know-how are essential requirements for future growth. To retain the best-qualified talents in the long-run employee aspirations have to be aligned with the needs of the company. Targeted personnel development and effective knowledge management are crucial in optimizing the value creation contribution of human capital.

Investor Communication

Any stock quote can be interpreted as the weighted average of the analysts' and investors' expectation scenarios. A sustained value creation effort, therefore, requires a credible and transparent *equity story*. Visibility on the capital market is a basic requirement for successful communication with investors. Credibility is gained through ambitious but realistic growth targets. The precondition is that they are consistently met, or, ideally, exceeded. A consistent fulfillment of short-term targets is of fundamental importance to support long-term growth perspectives.

The optimization of individual elements of the organizational foundation is necessary but not sufficient to increase and sustain shareholder value. Rather, this foundation must be brought into an overall balance. An evaluation of the efficient use of the elements may be guided by the following key questions:

Key Questions: Organizational Foundation

➤ Are all parts of the company fairly valued on the capital market, or could a portfolio optimization reduce potentially existing conglomerate discounts?

➤ Are the processes efficiently defined and scaleable? Are competencies and responsibilities clearly defined?

➤ Is there a clear understanding of the company's position within the industry's value chain?

➤ Are there conceivable approaches for lowering capital costs? How can additional company growth be financed?

➤ Is there a good balance of creativity potential and implementation know-how in the workforce?

➤ How high is the employee turnover? Are the best employees recruited, retained and developed?

➤ Do institutions exist to maintain contact with former employees and to utilize this external network constructively?

➤ Are there efficient and continuous communications with the investors? Is the picture conveyed to the capital markets realistic and current?

When implementing the House of Value Creation at a company, it makes a difference whether the company has only one business segment or whether it is a conglomerate. Direct application of the model is possible for companies with only one business segment. For conglomerates it is useful first to separate the company into business units or lines of business in order to take the unique requirements of the respective industries into account.

The following chapters describe the practical application of the House of Value Creation's elements.

3

Company Vision & Aspirations: The Roof of the House

The company vision defines the roof of the house. This chapter therefore focuses on the significance of this vision for increasing company value. We will demonstrate how heterogeneous aspirations of the senior management can be aligned with the company vision.

3.1 Clear Visions are a Prerequisite for Value Creation

Formulation of the Vision

The vision describes the company's core business, its strategic focus, and its future orientation. It could be formulated on an abstract level: "We intend to shape the future of e-Business." "We intend to produce the world's best communication equipment." "We want to grow profitably." More concrete goal statements are also conceivable: "With high quality products and increasing customer orientation, we intend to double our sales by 2006 and achieve a return on sales of greater than ten percent."

Since the entire company is guided by the vision, the statement should be both challenging and attainable. Unrealistic projections ("despite the current recession, we will double our sales annually" or "we will grow by 50 percent a year even without major investments") are not helpful and quickly lead to a serious loss of credibility among investors and frustration among employees. The examples of NEC and Sun Microsystems in figure 3-1 illustrate the positive impact of more abstract visions on the future development of a given company.

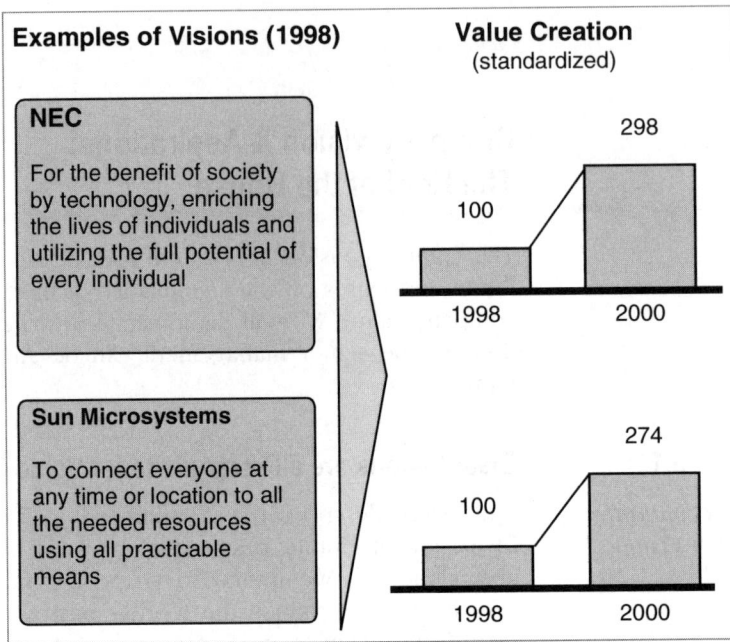

Figure 3-1:   Correlation between the visions and achievement at NEC and Sun Microsystems

Conglomerates in particular often have difficulties in defining their vision with sufficient specificity. General Electric dealt with this problem in an elegant manner by simply stating that its corporate goal was to become the No. 1 or No. 2 player in each market in which the company was active. Such a statement does not restrict any manager while at the same time focussing them on a comprehensive and generally applicable corporate goal.

Dynamics of the Vision   As far as possible, company visions should remain constant, so that the employees can identify with them. On the other hand, visions must be adaptable to significant market changes. In some instances, turning away from a business model entirely and overthrowing the existing vision statement completely was necessary in order to continue to deliver high value to shareholders. Intel (from memory chips to CPUs), Nokia (from pulp and paper to mobile phones) and Corning (from dishes and glassware to fiber optics) are just three examples.

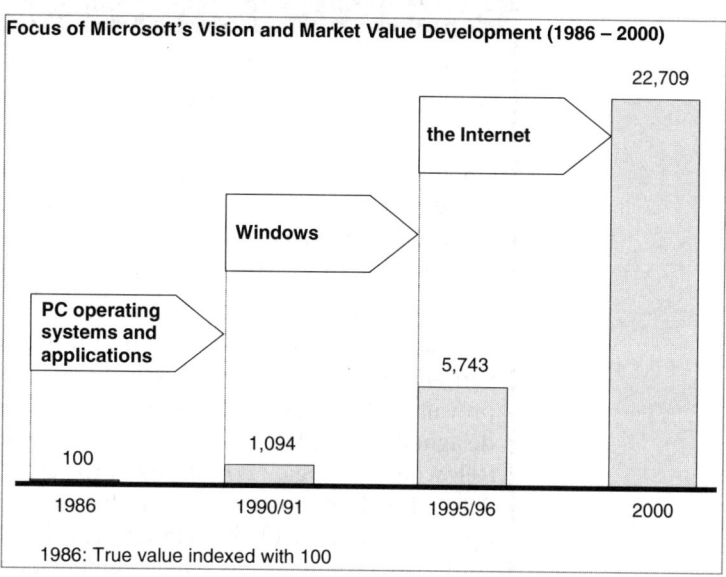

Focus of Microsoft's Vision and Market Value Development (1986 – 2000)

22,709

the Internet

Windows

PC operating
systems and
applications

5,743

1,094

100

1986   1990/91   1995/96   2000

1986: True value indexed with 100

Figure 3-2:   Vision and value creation at Microsoft

Microsoft's vision also changed with general industry trends. Within ten years the company successfully shifted its focus and expanded its vision from the production of application software to the offering of leading Internet-based solutions (figure 3-2).

Corporate Goals and Strategy

Since corporations are artificial entities, they cannot strictly speaking have their own goals. Corporate goals therefore can rather be viewed as the aggregation of all the goals of the individual stakeholders.

Different stakeholders naturally have different goals: shareholders expect suitable returns on their invested capital; employees expect attractive and secure jobs; and customers desire certain types and qualities of products. It follows that these heterogeneous goals and claims need to be smoothly integrated into a set of corporate goals valid and acceptable to all stakeholders. Within the space spanned by these goals, the company has to state how it intends to sustainably generate value. Based on that, strategies to reach the company goals are then derived and communicated in order to generate concrete guidelines for the daily business.

Box 2:   And What About the Strategy?

Clearly, the strategic direction and the ability to continuously adapt the strategy to changes are decisive success factors for a company.

On the one hand, the House of Value Creation views the process of strategy development as an implicit component of the development of vision and aspirations. In terms of the Cultural School of Mintzberg, this is a collective process on the executive level driven by visions and projected targets.

On the other hand, we have found through an evaluation of various project experiences that there is often a tactical gap prevailing between the corporate strategy and the programs designed to implement it – often this holds especially true following strategic consulting projects (Linn, Accenture, 2001). It is from this tactical gap that the key elements for the operationalization of the strategy can be derived. These include, for example, customer-segmentation strategies, channel strategies, product strategies, or manufacturing strategies. This also includes the organizational levers required for their effectiveness, such as incentive systems or human resource strategies. The tactical gap can be closed by implementation strategies, which derive from the individual components of the four levels of the House of Value Creation. In sum, these sub-strategies then make up the overall corporate strategy. Thus, strategy development is not considered a separate process but rather as an integral component of the individual elements of the House of Value Creation.

Aspirations as Operationalization of the Vision on the Individual Level

Individual Aspirations

Aspirations are the objectives of individuals and they affect both institutional as well as individual value contributions. Examples of aspirations could be: "our product should become so good that it will dominate the market," or "my department should be able to reduce costs by 20 percent within a year," or "I see myself as a creative force for change who links his own personal success with that of the project."

Naturally, these individual aspirations are not always congruent with the corporate vision, particularly if the latter appears to be too aggressive or not applicable within the individual's own area of activity. Therefore, the actual or apparent barriers to pursuing and attaining the corporate goals must be analyzed precisely in order to develop measures to overcome them. It is particularly critical to make sure that all members of key management pursue a uniform and realistic value creation goal. The alignment of these individual aspirations in accordance with the corporate goals in the real world generally involves a four-stage process, which is outlined in Chapter 3.2.

Alignment Towards Value Creation

The overall alignment of the company towards value creation can only succeed if all the relevant parameters are coordinated. These parameters in particular include: the (generally externally determined) valuation target, the goals of senior management as well as the organizational and processual prerequisites of a company.

In general, purely quantitative valuation targets are set externally (investor expectations) or are derived from competitor benchmarks. The objectives of individual executives cannot be so easily determined and are often quite heterogeneous. Often, dissension exists, at least implicitly, regarding key corporate goals and strategic foci.

Typical differences of opinion may exist along the following strategic questions: Do we seek market leadership and global expansion or would we rather increase our value by raising profitability? Should we continue to grow using our present product line or do we need to enter new business areas? Can a sales profit margin of ten percent or more be attained within two years?

Even organizational issues do not always generate consensus. Questions such as the following arise: Are we directing our efforts based on the appropriate key performance indicators (KPI)? Have our individual and institutional objectives been properly harmonized and can I actually impact the performance-based component of my remuneration package? Do our employees have the proper skills to develop new business? Is our organization as well as our corporate culture designed for rapid response and innovation?

Developing consensus among the management about the vision and goals of the company leadership is a critical success factor on the way to alignment towards value creation. It is this strategic direction from which the prioritization of activities for the lower levels derives.

3.2   Aligning Heterogeneous Aspirations Towards the Goal of Value Creation — A Pragmatic Guide

A Four-Step Model for Harmonizing Aspirations

Harmonizing the individual aspirations of the senior management is the prerequisite for the successful implementation of a value creation program. We therefore recommend a systematic process comprising four steps:

1. Goal oriented, structured individual interviews with the members of the senior management conducted by a neutral third party;

2. Anonymous quantitative and qualitative evaluation of the interviews in order to identify value creation potentials and eliminate any impediments for the realization of such potentials;

3. Performing plausibility analyses of the identified value creation potentials based on competitor benchmarks and existing business plans; as well as

4. Feedback and workshops for joint goal determination as well as the derivation of priorities along the levels of the House of Value Creation.

This approach not only aids in harmonizing individual objectives, but also improves communication and builds trust. It also generates important insights into the fundamental value creation parameters of the business. These are then prioritized and implemented along the other levels of the House of Value Creation. Examples of such parameters are the necessity of a growth program, improving the cost position, or optimizing the incentive and control systems.

3.2.1   Individual Interviews with Members of Senior Management

Step 1

First the decision makers responsible for value creation are identified. On the corporate level, these are typically the members of the senior management as well as the business unit heads (functions or divisions). Interviews of between one to two hours in duration are set up with each of these decision makers. Absolute confidentiality of the results by anonymizing the data has to be guaranteed in order to assure maximum frankness of the interviewee.

In the interview, both qualitative and quantitative aspirations are gathered. Examples of quantitative aspirations include sales, market share, profitability, or market valuation. Qualitative factors, for example, include thoughts regarding technology and product strategies. The target variables specified must be accompanied by a time horizon ("Increase sales 50 percent by the end of 2003") to permit comparability with the aspirations of other members of senior management. In the case of quantitative targets, the underlying projected market and competitive parameters should be spelled out. Thus, for example, the sales projections for telecommunication suppliers are presently heavily impacted by the expectations of the general market success of UMTS. Personal aspirations can best be determined if individuals are offered maximum space for decision-making as part of the interview. The starting point to determine such aspirations, therefore, are questions like: "What would you set as goals if you were Chairman of the Board and had total freedom to decide?"

In the second section of the interview, the interviewees are brought back down to (often harsh) reality. The questions now read: "In today's situation, what works against the realization of those ambitious targets?" Barriers generally specified include inadequate investment, insufficient skills (e.g., in Research and Development), organizational barriers (e.g., low willingness to make decisions, dilution of resources) or inappropriate, growth-inhibiting metrics.

The third section of the interview concentrates on the development of measures needed to overcome the value creation barriers identified in Part 2. In line with the structural logic of the House of Value Creation, the following questions arise: "You consider the control variables as impediments – what changes would you implement in order better to realize your goals as stated in Part 1?" Following a brainstorming and structuring phase, the interviewee should prioritize his proposed measures: "If you were Chairman of the Board, name the three most important measures which you would undertake immediately."

In the fourth section of the interview, the measures, which had been proposed in order to realize the company goals are lined up against the personal goals of the interviewee. The objective here is to clarify the individual´s own perception of his or her role within the company. Typical questions during this section are: "From your immediate perspective, what generates the biggest contribution to value creation?" Or, for example, "How do you see your position in this company in three to five years?"

It is best to conclude an interview with a short iterative loop: "You have only identified a few areas of development. Can you think of anything else?" Mutual feedback is also helpful: "Did you find this discussion useful?" Confidentiality should be assured once again at the conclusion and the next steps of this program or project at hand should be explained.

3.2.2 Qualitative and Quantitative Evaluation

Step 2

In Step 2, the growth and profitability expectations – the two main determinants of company value – are evaluated quantitatively and qualitatively. The numbers from the business plan serve as the reference points. Target values for a short-term horizon, for example, the next three to five years, can generally be estimated realistically by the interviewees. For longer-term estimates, the individual expectations regarding the perpetual growth and profitability rates also need to be collected and compared in order to determine the *terminal value*.

In the example in figure 3-3, expectations for the period 2001 through 2006 for GreatValue company are plotted. On the horizontal axis, the projected return on sales (ROS) as a measure of profitability in the year 2006 is plotted. The vertical axis plots the average annual growth in sales from 2001 through 2006. The existing business plan is based on a sales margin of 13.0 percent in 2006 and a 2.3 percent average growth until 2006. In figure 3-3 the business plan targets are shown by the broken lines.

The individual interview results are denoted by small circles. The triangle represents the arithmetic average of the interview results. The results plotted in this chart are quite typical, based on the findings of the authors.

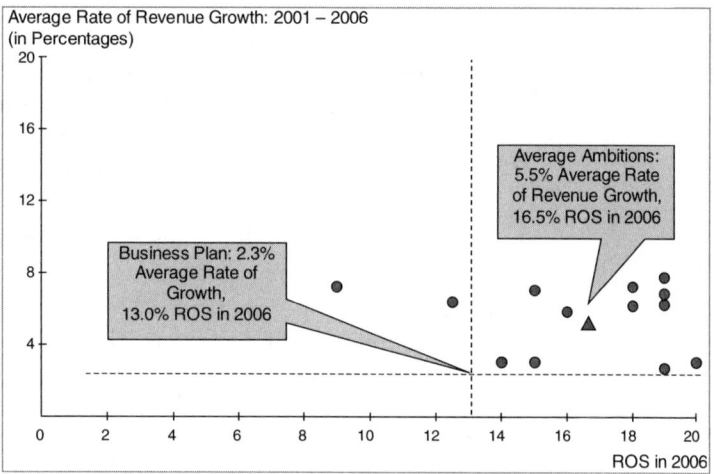

Figure 3-3:   Individual aspirations vs. business plan at GreatValue, Inc.

Three observations can now be made:

1. The interview results display a significant spread. This indicates that very divergent perceptions of the "attainable" prevail among the interviewed members of the senior management. This naturally also influences the personal perceptions regarding the necessary strategic foci (e.g., meaning focus on growth versus focus on profitability).

2. The average values of the individual aspirations, both in terms of profitability (16.5 percent) and growth rate (5.5 percent) were significantly higher than those stated in the business plan (13.0 and 2.3 percent, respectively). In sum, the management felt it could achieve a good deal more than the business plan laid out.

3. All of the interviewees consider a higher growth ratio be desirable. With only two exceptions, the interviewees also assume that a higher margin on sales is attainable. This would indicate that the growth dynamics of the company does not yet exhaust the market potential so that there is still room for improvement. Presumably, further potential also exists in terms of improving profitability.

The barriers to value creation as well as the measures needed to overcome them can be evaluated quantitatively. Figure 3-4 depicts such a barrier analysis (for the necessary measures, the illustration would be structurally the same). Each mentioning of a barrier is counted and the comments of the interviewees are structured and arranged accordingly by the evaluators. This is why the interview should be conducted along at structural logic as defined by the House of Value Creation.

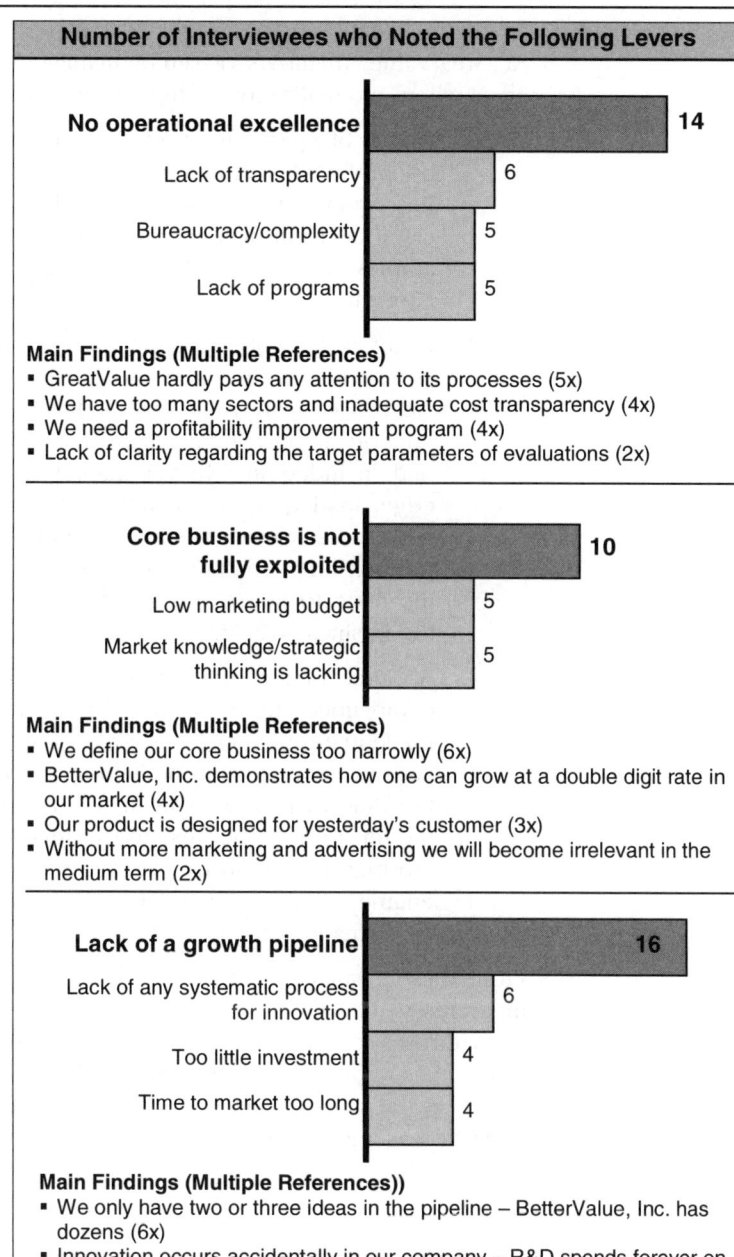

Number of Interviewees who Noted the Following Levers

No operational excellence — 14
Lack of transparency — 6
Bureaucracy/complexity — 5
Lack of programs — 5

Main Findings (Multiple References)
- GreatValue hardly pays any attention to its processes (5x)
- We have too many sectors and inadequate cost transparency (4x)
- We need a profitability improvement program (4x)
- Lack of clarity regarding the target parameters of evaluations (2x)

Core business is not fully exploited — 10
Low marketing budget — 5
Market knowledge/strategic thinking is lacking — 5

Main Findings (Multiple References)
- We define our core business too narrowly (6x)
- BetterValue, Inc. demonstrates how one can grow at a double digit rate in our market (4x)
- Our product is designed for yesterday's customer (3x)
- Without more marketing and advertising we will become irrelevant in the medium term (2x)

Lack of a growth pipeline — 16
Lack of any systematic process for innovation — 6
Too little investment — 4
Time to market too long — 4

Main Findings (Multiple References))
- We only have two or three ideas in the pipeline – BetterValue, Inc. has dozens (6x)
- Innovation occurs accidentally in our company – R&D spends forever on something and hopes that marketing will find a market for it (5x)
- Every year I have 5 projects which are stopped due to a lack of money – BetterValue, Inc. has implemented precisely my ideas (4x)

Figure 3-4: GreatValue: Evaluating the barriers

At GreatValue, members of senior management identified three areas where the realization of high aspirations were impeded:

1. A lack of operating excellence, meaning a lack of focus on profitability along the processes of the value chain. The interviewees named complex processes, bureaucratic decision-making and a lack of control tools as examples and demanded a systematic profitability enhancement program.

2. Insufficient exploitation of the core business. From the perspective of the interviewees, GreatValue could generate a significantly greater market share and thus more sales in its traditional business. Main deficiencies specified included insufficient market and customer knowledge, inadequate marketing and sales budgets and targeting the wrong customer segments with the wrong product lines. In sum, the interviewees clearly identified the need to improve internal capabilities and to realign the business with respect to the go-to-market strategy.

3. Absence of any growth perspectives. The interviewees complained about the lack of new product development, inadequate financial resources for new developments and an unsystematic innovation process, which delayed the time to market. This area was repeatedly ranked as the top priority, because in dynamic businesses - such as the high tech industry - a poorly functioning innovation engine would have fatal consequences on the future performance of the company.

From these results – particularly from the prioritization of the measures to be taken – it is possible to directly sketch out the main thrusts of a value creation program. In the following two steps, these thrusts will be subjected to plausibility analyses based on the market- and competitive situation and then institutionalized as a joint management agenda.

3.2.3 Plausibility Analysis of the Results

Step 3

In Step 3 the interview results are analyzed in the context of the market situation, the competition, and the company´s own business plan. The goal is to determine whether the aspirations derived from the interviews are plausible or whether they reflect unrealistic pipe dreams. For instance, it would clearly be unrealistic to expect double-digit annual growth rates in a stagnating and concentrated market if the company already controls a 30 percent market share. Step 3 therefore depends on the overall environment to a large degree and thus has to be performed independently for each business unit.

The comparison with the business plan will be illustrated along the example of GreatValue, Inc.: in figure 3-5, the overall corporate business plan covering the period from 2001 to 2006 is plotted along the sales and sales margin axes:

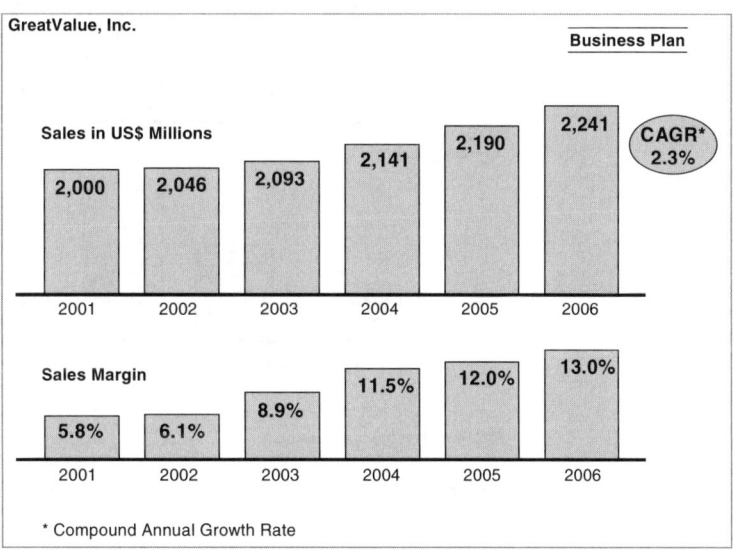

Figure 3-5:   Business plan with 2.3 percent sales growth and 13.0 percent profitability

The business plan assumes a compound annual growth (CAGR) in sales of 2.3 percent. The margin is expected to increases continually from 5.8 percent in 2001 up to 13.0 percent in 2006.

The interview results (average values) are plotted as a range on the time axis in figure 3-6. Already in 2004, a major revenue gap with a volume of US$522 million opens up between the aspirations of the senior management and the business plan. This gap widens further to approximately US$745 million by 2006. In terms of the sales margin, there is also a discrepancy of 5 percent, which by 2006 declines slightly to 3.5 percent.

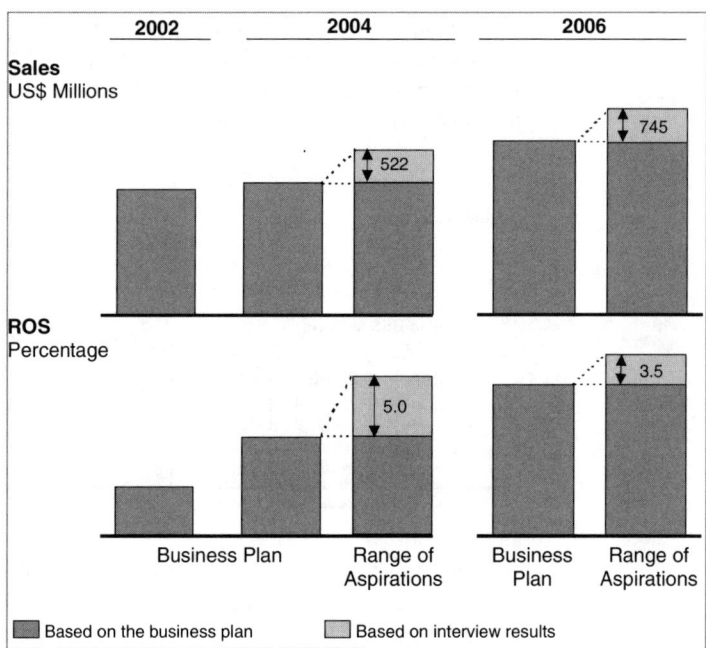

Figure 3-6:   Comparison of business plan and aspirations of senior management

The difference between senior management aspirations and the business plan can also be expressed in terms of company value. Figure 3-7 shows various valuations of GreatValue, Inc. The leftmost column shows the actual capital market value - US$1.5 billion.

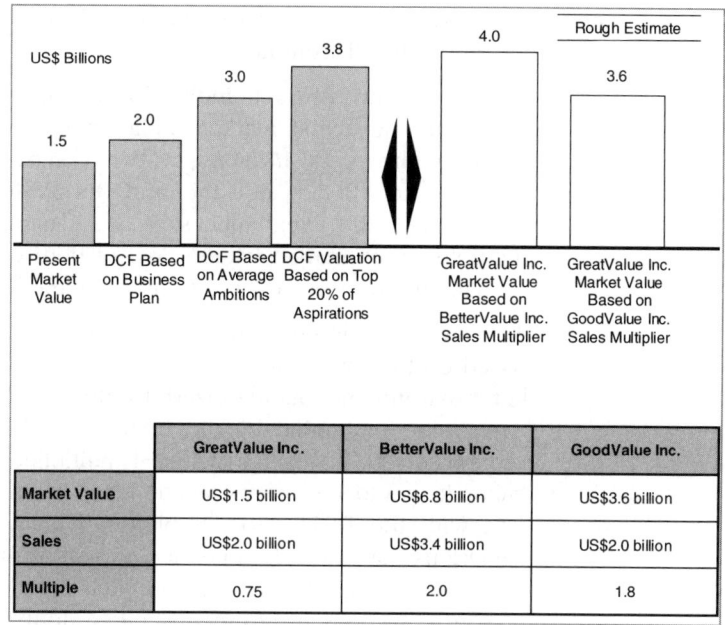

Figure 3-7: Market value of GreatValue compared to competitors

Based on the business plan, the calculated value would be approximately US$2.0 billion (using the *Discounted Cash Flow* methodology with the assumptions for the Terminal Value taken from the business plan; see appendix B). This figure is plotted in the second column. Thus, if the business plan is considered realistic, then clearly there is a communication or confidence gap between the goals communicated in the business plan and the market valuation.

The third column represents the Discounted Cash Flow (DCF) evaluation of the average quantitative aspirations (growth and sales margin targets) identified during the interviews. At approximately US$3 billion, it is significantly higher than the corresponding valuation based on the business plan. The realization of the aspirations would therefore increase the company value by roughly an additional US$1.0 billion.

For comparative purposes, the maximum value based on the highest aspirations was also calculated. The top 20 percent of responses, representing the most ambitious aspirations, produced a company value of US$3.8 billion – 2.5 times the present market valuation (fourth column).

The plausibility of such individual valuations can be checked based on the actual market valuation of competitors. For that purpose, the *Sales Multiples* of two competitors listed on the stock market (BetterValue Inc. and GoodValue Inc.) were computed. Then the GreatValue sales were multiplied by the sales multiple (stock market value/sales) of both competitors in order to make the market capitalizations comparable.

As shown in columns 5 and 6 of figure 3-7, the valuation targets based on the interviews are by no means unrealistic in this market environment. Standardized to the sales, the valuation of GreatValue multiplied by the BetterValue factor would be approximately US$4.0 billion while if multiplied by the GoodValue factor it would be approximately US$3.6 billion. This is consistent with the US$3.0 to 3.8 billion, which emerged as target valuations based on the interviews. In short, there are definitely players in the market who attain such high valuations. For that reason, it makes sense to incorporate a doubling of the valuation to at least US$3.0 billion as a common goal into the management workshops described in step 4.

3.2.4 Feedback and Goal Specification Workshops

Step 4

In the final step, the decision-makers are sworn to the vision. Priorities are set for strategic thrusts to increase the valuation and a process is initiated to identify the required value creation managers (figure 3-8).

Experience shows that the interviewees should get together for that purpose in a joint workshop of at least half-day duration – ideally outside the usual workplace. The moderator (generally the evaluator of the interviews) first presents the results of the interviews. Then, in the context of the items relevant for the decisions to be made (quantitative goals, prioritizing of the main thrusts), a discussion is held and a secret voting is conducted. In this way, personal and hierarchic reservations can be avoided when formulating the joint commitment.

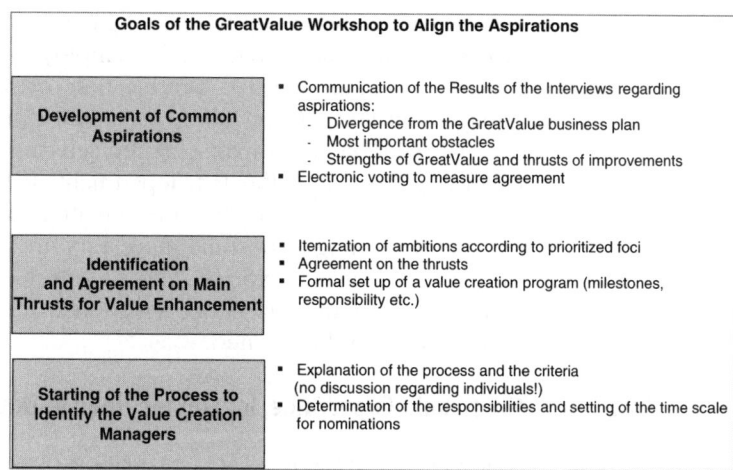

Figure 3-8: Goals of the GreatValue workshop to align aspirations

Interestingly, in this process the results of the efforts to form a joint commitment usually converge relatively quickly despite the variances expressed in the interviews. This is due to the fact that the individual perceptions are put into a more relative perspective when lined up against the market environment, business plans and inputs of colleagues.

In order to implement the incipient initiatives and the correlated, generally radical changes, a sufficient number of value creation managers need to be identified. Their tasks can be defined along three dimensions:

1. Implementation of the value creation initiatives across functions and hierarchies

2. Support for the commitment to operational excellence and value orientation at all levels of the organization and

3. Assuring the sustainability of the value creation program by institutionalization.

The ensuing selection process is rather complex. Given the high value creation potential of approximately US$1.5 billion in GreatValue's case, the careful selection of the key employees for the implementation of the initiatives is of utmost significance.

First, specific task lists need to be developed for the value creation managers based upon the change requirements and key success factors. Thereafter, suitable employees must be identified through a comprehensive preselection cycle on the defined management levels (particularly at middle management). Eventually, the value creation managers are selected jointly by the senior management and the HR department. Following a comparison between the need for value creation managers and their availability, it might become necessary to prioritize the tasks and/or hire outside experts. Finally, specific tasks can be assigned to the individual value creation managers and performance metrics can be defined.

3.2.5 The Result: GreatValue, Inc. on the Road to Doubling its Valuation

Following the interviews and evaluation phase, the management of GreatValue jointly agreed on the following results:

1. GreatValue, Inc. sets the goals of doubling its market value (currently US$1.5 billion) within 24 months.

2. The goals specified in the business plan have to be significantly exceeded. The targets are a higher growth rate (at least 5.5 percent instead of the planned 2.3 percent) and a higher level of profitability (5 percent more in 2004 than currently planned). The business plan is to be adjusted accordingly.

3. A systematic value enhancement program is to be initiated. The key thrusts and thus also the success parameters of the program will be: "Improving Operating Excellence," "Full Exploitation of the Core Business," and "Growth and Innovation Campaign?."

4. The status of the program will be reviewed regularly and, if necessary, adjusted to changed circumstances (for example, through a staff position for value management).

5. The senior management nominates a number of value creation managers (three to five per main initiative) who will supervise the specific initiative on a full time basis.

With these unified aspirations and action plan, GreatValue is now well equipped to detail? the specific thrusts and to embark on the journey to value creation. The roof has been completed. Now the pillars, growth and profitability, have to be erected in order to set the roof on a solid structure.

4 Value Creation Levers: The Supporting Pillars

Two strong pillars need to support the roof of the house. In our framework, these two pillars are represented by the two value creation levers profitability and growth. Their purpose is to implement the vision and increase the company value in a directly quantifiable way. To do this, the value gap to best-in-class needs to be broken down along the dimensions profitability and growth first. Then, a suitable combination of the two value creation levers can close these two gaps.

4.1 Managing for Profitability or Growth?

Only profitable growth can create sustainable value – and the prerequisite is a balance between the value creation levers. Their relative influence on the valuation of a company varies with the current market value and the valuation level of the industry as a whole. Senior management thus has to analyze the initial position first before evaluating and balancing the impact of the value creation levers. The company's current market value and the valuation level of its industry also need to be taken into account when adjusting the metrics and incentive systems in the course of a value creation program. This is explained in more detail in chapter 5.

4.1.1 Significance of the Levers as a Function of the Company Situation

This chapter first illustrates the impact of the current valuation level on the relative significance of the two value creation levers, using a concrete example from the high-tech industry.

Figure 4-1 shows the market valuations of Sun Microsystems und Apple Computers, normalized by means of the sales multiple. Mathematically, this representation follows the DCF-method which is explained in more detail in appendix B. Straight lines represent isoquants of all the combinations of perpetual growth and perpetual profitability, which result in the same relative valuation, i.e., the same sales multiple.

For better illustration the model has been simplified here to assume the terminal value accounts for the total company value
instead of just 80 to 90 percent, which is common in dynamic
industries.

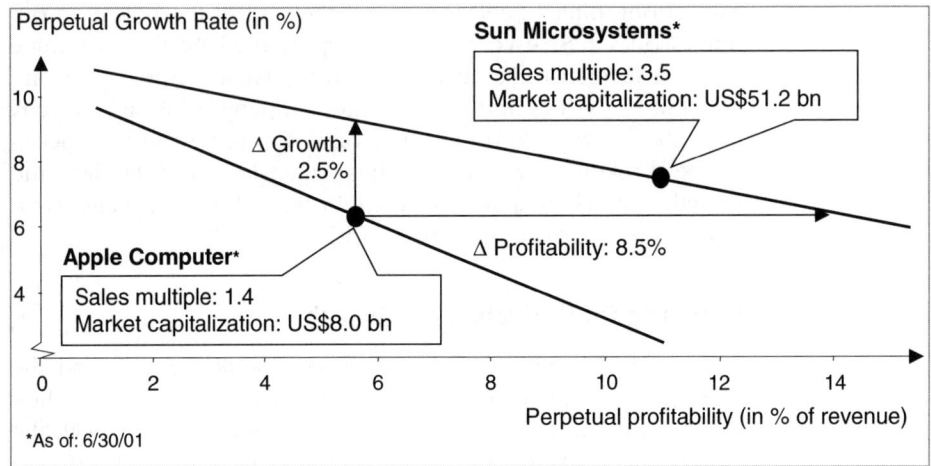

Figure 4-1:   Dependence of the impact of the value creation levers on the
capital market valuation

Depending on the valuation level, the lines have different slopes:
Higher valuations result in flatter isoquants. Changes in the company valuation cause a shift onto a different value isoquant. For
instance, an increase of the valuation results in a shift upward
and to the right. In practice, a change in valuation will always be
based on a simultaneous change of both profitability and
growth, to varying degrees. In order to illustrate the relative sensitivity, however, the migration path needs to be decomposed
into the movements along the individual axes. Thus, both levers
need to be considered *ceteris paribus*. Depending on the initial
position, i.e., the current combination of growth, profitability,
and slope of the isoquant, the migration paths along the axes
have different length.

Sensitivity of the
Value Creation
Levers

The points in figure 4-1 denote the current positions of Sun Microsystems and Apple Computer on their respective isoquants. In order to reach Sun's higher valuation level, Apple can employ one or a combination of both value creation levers. Figure 4-1 indicates that increasing the perpetual growth rate by 2.5 percentage points, all else equal, would be sufficient to close the valuation gap. However, if Apple were to rely only on improving profitability, this lever would have to be pushed by 8.5 percentage points to achieve the same effect.

While in this example the vertical path (i.e., via increasing growth) to a higher value isoquant is shorter, at least in absolute percentage points, a horizontal movement (i.e., via improving profitability) is often more attractive. This is especially true for relatively low valued companies in industries with low overall valuation levels. In such industries, the increase of the perpetual profitability, which is required to move to a higher value isoquant, would be lower than in the example in figure 4-1.

Also, in case of a relatively low valued company it should be considered that the potential for increasing the perpetual growth rate is quite limited. Industries with low valuations are usually less dynamic and further along the typical industry life cycle. In those industries it is usually easier to substantially improve profitability than to raise the perpetual growth rate by even a fraction of a percentage point.

Since no company can grow faster than its industry or the overall economy forever, growth rates cannot realistically grow overproportionally for an extended period of time. This aspect also needs to be taken into account with relatively high-valued companies. The capital markets will view excessive growth targets critically, if the company already controls a significant share of the market. Failure to meet the targets would then surely result in a significant downward correction of the market valuation. There are a sufficient number of examples for such corrections. Not just start-ups, but also large, well-established companies like Nortel Networks or Lucent have suffered painful declines of their valuations because they have disappointed investor expectations regarding growth.

In sum, the sensitivities of the two value creation levers growth and profitability differ depending on the current value of the company as well as on the overall industry valuation. Thus the company leadership needs to analyze what increases of the individual levers are realistic and which of the two value creation levers will generate higher value growth in the short-term. However, in the long run, only a sound balance of both value creation levers will ensure sustainable value creation.

Value benchmarking allows further analysis of the dependence of company valuation on growth and profitability. Along the example of GreatValue, Inc., this instrument will now be described in detail.

4.1.2 Value-Benchmarking Against the Best-in-Class – Not Just Profitability Counts

In order to extract as much information as possible from this analysis, the company value is not calculated macroscopically, but through a sum-of-the-parts calculation, where the values of the individual business units are aggregated. The appropriate valuation of the three business units (G1, G2 and G3) is calculated according to the DCF-method and substantiated by sales- and profit multiples. The basis of this calculation are the business plans of the individual business units, which can be projected further into the future if necessary. A forecast period of three years is reasonable in dynamic industries. In established industries the business plan could be extrapolated over a period of five or more years into the future. A longer forecast period reduces that portion of the total value, which is accounted for by the terminal value. By definition, the terminal value makes up that part of a company's valuation which lies beyond the explicit forecast period. Selecting a longer forecast period is appropriate only if that forecast is more exact than the estimate of the constant, perpetual figures.

The overall market growth serves as an approximation for the perpetual growth rate. In addition to the perpetual profitability it constitutes a determining factor for the terminal value. The cost of capital is derived from figures for the comparable business units of competitors, or based on analysts' estimates. In order to increase plausibility, the results of the DCF-analysis are compared to those of additional analyses, such as the separation of total company value into revenue- and profit shares, or external comparisons via revenue and profit multiples.

In case of our example company, GreatValue, Inc., a sensitivity analysis is conducted based on the business plans for the three selected business units. This analysis indicates how sensitively the value of the selected business units reacts to changes in key parameters. The following key parameters used here have proven useful proxies for the purpose of determining the relative significance of the value creation levers:

- Compound annual revenue growth during the observation period relative to the base value;

- EBIT-margin at the end of the observation period;

- Marginal increase of the company value for a one percent revenue increase;

- Marginal increase of the company value for a one percent increase in profit margin (ΔEBIT).

The results of this analysis now provide a basis for a comparison of the business units with significant competitors (figure 4-2). Data on these benchmark competitors can often be derived from their financial statements. Should this prove difficult, the individual parameters will have to be collected from diverse sources and checked for plausibility.

For instance, the revenue breakdown by individual business unit can be approximated from market shares, production volume, and similar data. By contrast, assessing the profitability of the individual competing sectors often proves to be more difficult.

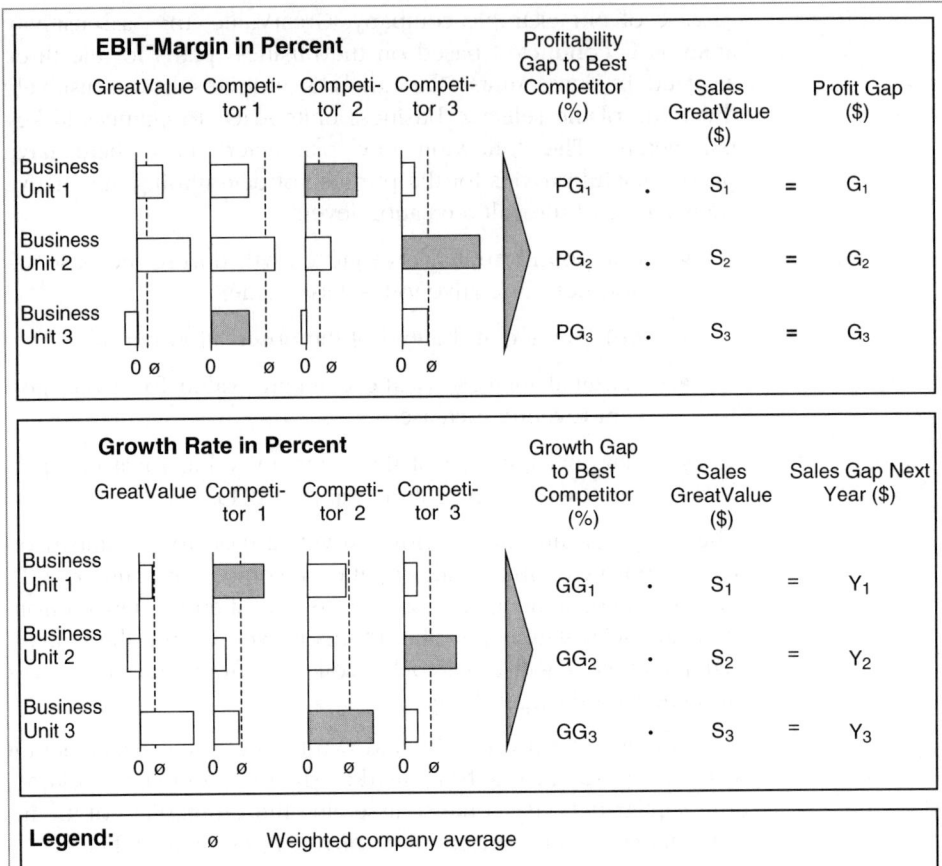

Figure 4-2: Schematic graph of the benchmarking between GreatValue's business units vs. those of main competitors

For the evaluation of profitability of a particular business unit, the following process has proved to be successful: the annual report of a competitor usually lists overall profitability of a competitor. If figures are provided referring to the profitability of some of the individual business units, or if related analyst opinions are available, some values can be taken as fixed points. The weighted average (based on the estimated sales distribution) of the profitability of the remaining business units must then equal the overall company profitability. This method is similar to a puzzle where numerous employees, e.g., strategic planners, controllers and marketing experts, can provide valuable input. They, too, may derive benefits for their own departments from such a procedure.

The value gaps between the business units of the examined company and their respective most successful competitors can be separated into profit gaps and sales gaps. In the context of such a value benchmarking, the best-in-class can be thought of as a hybrid, which is assembled from the most successful business units of different competitor companies. The absolute extent of the value gap is determined in comparison to this construction.

The profit gap is the difference between the profitability of the analyzed business unit and that of the most profitable competitor. Analogously, the sales gap is the difference between the growth of the own business unit versus the business unit of the fastest growing competitor. The analysis shows to what extent both gaps are finally responsible for the total value gap. In this way, foci for the necessary improvement measures within the analyzed business units can be derived.

4.1.3 Integration of Both Pillars into a Successful Value Creation Program

After decomposing the valuation gap into the two components, *potential for profitability improvement* and *growth potential,* both now need to be optimized by means of a comprehensive value creation program.

Profitability Improvement Program

A typical profitability improvement program is graphically represented in figure 4-3. Such a program typically has a long-term focus. The implementation can take from several months to several years.

The as-is analysis at GreatValue, Inc. has revealed a significant profitability gap compared to competitors and an urgent need to act. Internal analyses and the evaluation of the value benchmark have shown that costs can be reduced drastically through reduction of over-capacities and improved efficiency.

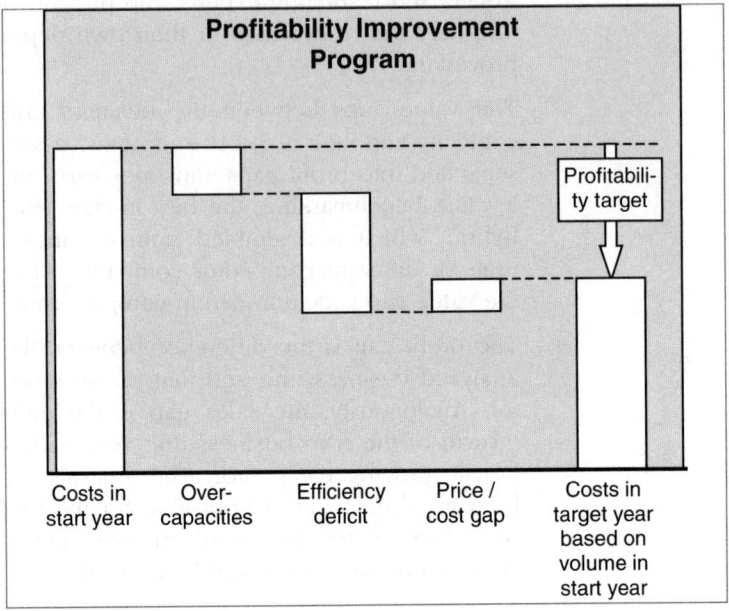

Figure 4-3:   First phase of a value creation program

When determining the target values, any possible cost increases and potential price declines (price-cost gap) which are to be expected during the course of the profitability improvement program have to be taken into account. This dynamic adjustment of prices and costs typically increases the urgency to act even more. The target costs in the target year are then calculated based on the current (start-year) volume.

Based on the profitability target defined in the value creation program, a comprehensive profitability improvement program can now be designed. The process of planning and implementing such a programs is discussed in detail in chapter 4.2.

The second phase of a value creation program deals with growth (figure 4-4). Challenging yet realistic growth objectives have to be defined through the target year – both for the existing core businesses as well as for new business areas. Here it is important to maintain a suitable balance between the different growth horizons. Neglecting the development of new businesses in favor of expanding the existing ones has to be avoided – and vice versa. The value creation lever growth is discussed in chapter 4.3.

4.2   Increasing Profitability Through Cost Optimization

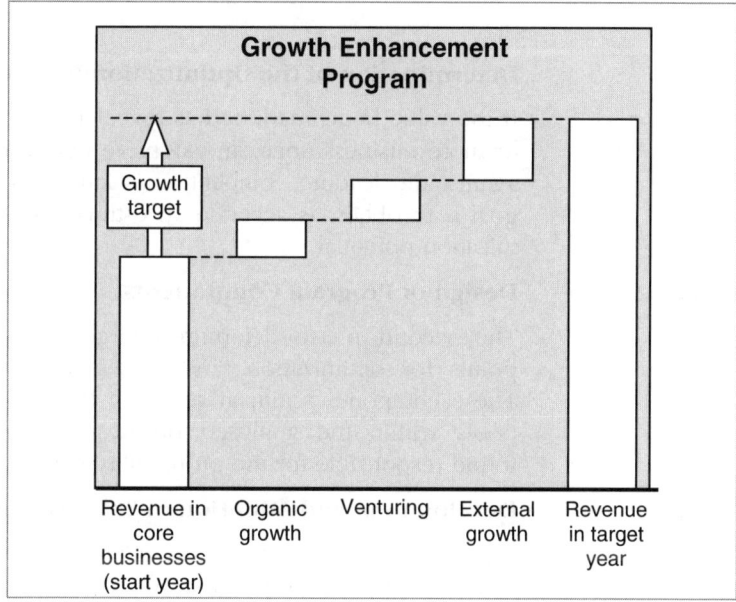

Figure 4-4:   Second phase of a value creation program

Through the targeted approach of a profitability improvement program, costs can be reduced and thus profitability increased. Such a program is organized in four phases (figure 4-5):

Phase 1

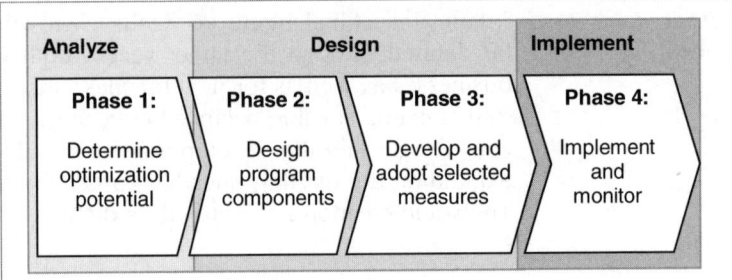

Figure 4-5: Phases of the profitability improvement program

Determination of the Optimization Potential

If the value benchmarking has revealed a profitability gap at one or more business units, an extensive performance benchmarking against the leading competitors is launched in phase one. The goal is to obtain as accurate an estimate as possible of the optimization potential.

Phase 2

Design of Program Components

The second phase determines target costs, identifies starting points for optimization, and draws up program components. These components aim at reducing the relevant functional cost pools within the analyzed business units, which have been found responsible for the profitability gap.

Phase 3

Development and Adoption of Selected Measures

After program components are in place, a comprehensive profitability improvement program is conducted, in which detailed cost reduction measures and initiatives are developed, adopted and authorized.

Phase 4

Implementation and Monitoring

These measures and initiatives are then implemented in phase four. The implementation of the entire program is continuously supervised and monitored as well as updated or adjusted to changes in business conditions.

These four phases will now be described in detail.

4.2.1   Determination of the Optimization Potential

Phase1

Improving the profitability of a company requires an in-depth understanding of costs and cost drivers. If the comparison with the leading competitors identifies a profitability gap, this gap needs to be validated and decomposed to the level of its causes. Thus, the next steps consist of a functional benchmarking and a detailed drill-down analysis (figure 4-6).

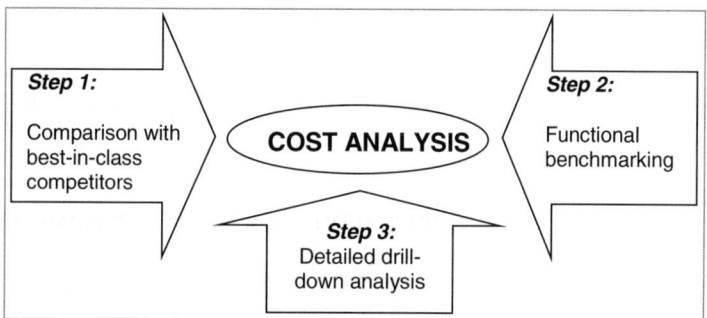

Figure 4-6:   Steps for detailing a profitability gap

For instance, a functional benchmarking could entail comparisons of key performance indicators of individual production areas or overhead ratios. Detailed drill-down analyses permit further validation of the top-down comparison with competitors.

4.2.1.1 Comparison With the Best-in-Class

Step 1

The comparability of reference values is a well-known problem in benchmarking. In order to allow for a meaningful comparison of global performance parameters such as profitability, productivity, and efficiency these figures need to be normalized (e.g., value added per employee, purchasing costs per unit volume etc).

Data about competitors can most often be compiled from annual reports, broker reports, or similar publications. In case the necessary data are not accessible through such standard sources, for instance in case of privately held companies with limited statutory disclosure requirements, the database has to be assembled from alternative sources. Examples here are press releases, interviews with former employees, expert assessments, or even reverse engineering of competitor's products. An active and direct involvement of the competitors in the benchmarking process can improve the quality of the database significantly. However, from a strategic point of view this often proves to be problematic.

For the top-down comparison of the company with the best-in-class it is often useful to compare the operating costs per unit produced. These costs can be computed by subtracting profit, taxes, and depreciation from sales. Because in practice product ranges are usually heterogeneous, the costs first need to be allocated to the products involved in the comparison. To ensure comparability, the cost structures need to be standardized and the costs need to be distributed to the respective cost pools. Figure 4-7 shows a decomposition of the global cost structure into personnel costs, capital costs, and material costs.

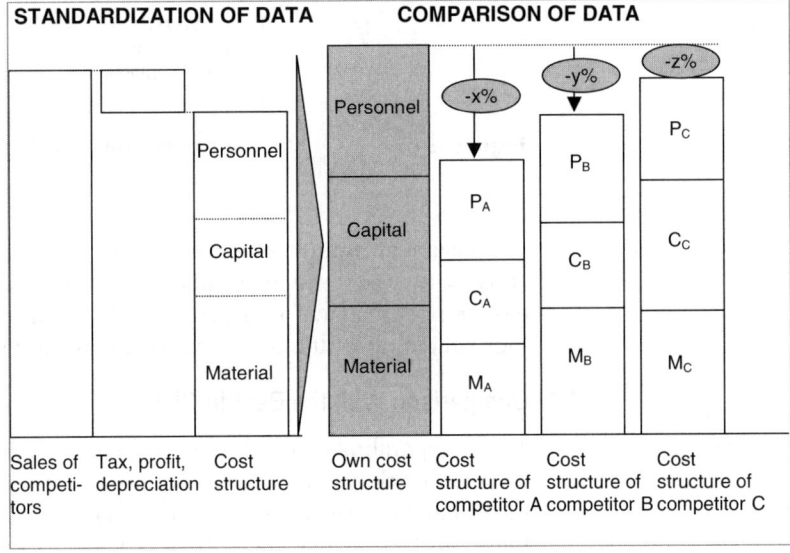

Figure 4-7: Top-down comparison of cost structures per unit produced

The profitability gap is determined by comparing the total costs, or the three costs pools per unit produced of the benchmarked company against the main competitors.

| *Analysis of the Cost Gaps* | An analysis of the cost drivers and cost pools reveals the cost gap of the company versus its competitors. These gaps are a first indicator for the value creation potential of the company. The depth of value added between the companies surveyed can cause shifts between personnel and material costs. In this case, adjustments need to be made here based on reasonable estimates before proceeding. Productivity measures thus need to take different depths of value added into account. It is therefore useful to compare the value added per employee (sales per employee minus procurement expenses per employee) rather than just sales per employee. |

Alternatively, the cost structure of the respective cost leader in each cost pool can be used as a benchmark. This provides information about the absolute potential for optimization, which ideally is attainable in each of the three cost pools. A weak point of this approach is the potentially different definition of the cost pools among the companies benchmarked, with overlaps leading to distorted results.

| *Benchmarking Matrix* | Because costs depend on a diverse array of factors, cost pools cannot simply be added and directly compared with competitors. Different countries might produce at different factor costs, or economies of scale may play a role in certain businesses with a high share fixed cost share. Thus, a benchmarking matrix (figure 4-8) is used for the normalization of the cost pools. For clarification purposes, the matrix in figure 4-8 has been filled with values taken from an actual consulting project. The values that relate the three cost pools to their main influencing factors are shown in brackets. Values less than one indicate a disadvantage in the benchmarked company's cost structure versus that of the competitor. |

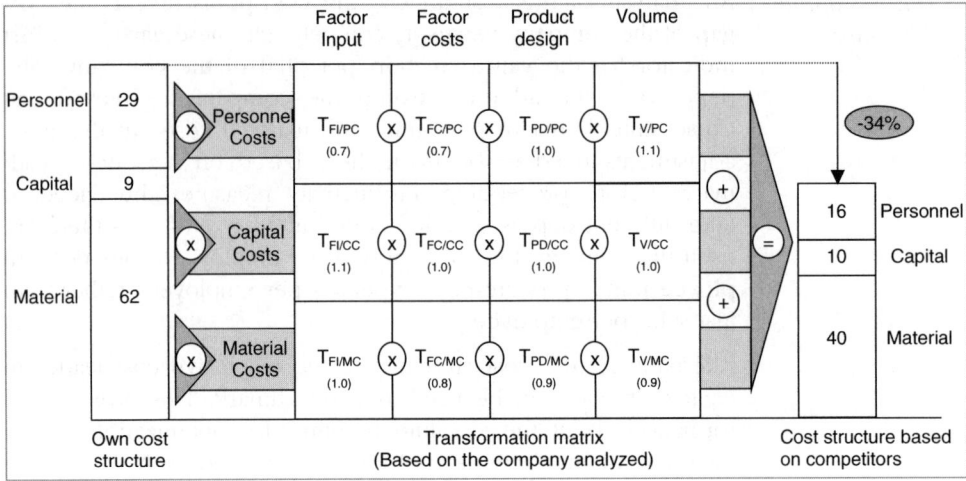

		Factor Input	Factor costs	Product design	Volume				
Personnel	29	x Personnel Costs	$T_{FI/PC}$ (0.7)	x $T_{FC/PC}$ (0.7)	x $T_{PD/PC}$ (1.0)	x $T_{V/PC}$ (1.1)		-34%	
Capital	9	x Capital Costs	$T_{FI/CC}$ (1.1)	x $T_{FC/CC}$ (1.0)	x $T_{PD/CC}$ (1.0)	x $T_{V/CC}$ (1.0)	+ =	16	Personnel
								10	Capital
Material	62	x Material Costs	$T_{FI/MC}$ (1.0)	x $T_{FC/MC}$ (0.8)	x $T_{PD/MC}$ (0.9)	x $T_{V/MC}$ (0.9)	+	40	Material
	Own cost structure		Transformation matrix (Based on the company analyzed)					Cost structure based on competitors	

Figure 4-8: Comparison of own cost structure to a competitor using the benchmarking matrix

- **Factor input** is a measure of productivity and efficiency. It describes a unit's performance per reference unit (e.g., hours needed to manufacture a car, amount of steel required per car).

- **Factor costs** quantify the costs per unit (e.g., personnel costs per hour, price per ton of steel).

- **Product design** summarizes factors resulting from a cost-efficient construction of the product (e.g., modular design, parts per product).

- **Volume** comprises effects that lead to decreasing unit costs, mainly in businesses with a substantial amount of fixed costs (e.g., additional cost reduction achieved by a two percent increase in capacity utilization).

In the following, the application of a benchmarking matrix will be illustrated using an example. In this example, only the personnel costs will be compared to those of an international competitor, which manufactures in a country with a lower wage level. This competitor, therefore, has a substantial factor cost advantage. In order to compare productivities, the costs need to be adjusted by this factor. Ideally, specific company values are used for this adjustment. If they are not available it may be possible to resort to statistics (box 3).

The following step analyzes the volume effect. Each cost pool is divided into fixed and variable costs. Both are then correlated with the number of units produced. The fixed cost degression yields another cost (dis)advantage, which can be computed by inserting the number of units produced by the competitor into the calculation of the benchmarked company's average cost per unit (fixed costs / number of units + variable cost per unit). Adjustments need to be made for different fixed cost shares, if necessary.

In addition to factor costs and volume, factor input and product design are also drivers of the cost gap. These too have to be standardized. In practice, factor input and product design are difficult to separate. The following logic might aid in this task:

1. Is there a discernible difference between the product designs? If not, then a separation of the two factors is unnecessary. The more complex a product and the deeper the value added, the higher the likelihood of a substantial difference. If there is indeed a difference in product design, a second question arises:

2. Can this cost difference be reasonably quantified, for instance, through reverse engineering? If quantification is not possible, a separation of the cost factors is nearly impossible. Nevertheless, the development of optimization measures should include the product design.

Box 3:   Productivities in Different Countries

Figure 4-9 shows the factor cost differences between several countries on a macroeconomic level as well as a comparison between the relative productivities. Such statistics are readily available from the statistical offices of individual countries or from international organizations, such as the *OECD*. According to the OECD statistics shown below, the hourly wage costs in Korea are US$5.60 compared to US$14 in Germany. However, the productivity difference of 62 percent in favor of the German worker also has to be taken into account. The benchmarking matrix normalizes personnel costs on the basis of equal productivity. As a result, the standardized work hour in Germany is even cheaper than in Korea. The normalization levels the enormous factor cost differences to a large extent – they end up relatively close together, in a range of US$14-18 (with the exception of France and Japan).

	Wages in US$/Hour[1]	Produc-tivity[2]	Standardized Work Hour in US$
Germany	**14.0**	**100%**	**14.0**
France	11.4	107%	10.7
Japan	21.8	92%	23.7
Canada	12.8	87%	14.7
Korea	5.6	38%[3]	14.7
UK	12.8	72%[3]	17.8
USA	20.0	117%	17.1

[1] Including Supplementary Costs, Source: UBS, Prices and Earnings 2000
[2] Value Added per Hour 1998, Source: OECD, World Bank 2001
[3] Value Added per Hour 1996, Source: OECD, World Bank 2001

Figure 4-9:   Comparison of wage levels and productivity of selected countries

Factor Input as a Measure of Productivity

After calculating the difference in product design, factor input remains as the residual value, representing the productivity. This variable is computed and can be calculated by solving the following equation for the factor input as the relative disadvantage:

Unit costs of the benchmark = own unit costs * relative disadvantage in factor input * relative disadvantage in factor costs * relative disadvantage in product design * relative disadvantage in volume

This relative disadvantage in factor input is the most important number in this entire analysis, because it constitutes a measure of the (hypothetical) optimization potential through an increase in productivity/ efficiency. It serves as a target value for the optimization program that will be set up in phases 2 and 3.

In practice, capital costs hardly differ between the benchmarked companies, as long as they all have equal access to capital markets.

Costs of services and material frequently represent significant cost drivers. In manufacturing industries (e.g. the automotive industry) they account for close to 90 percent of total costs due to the outsourcing of substantial parts of the manufacturing process. Material costs often vary significantly, because of the substantial impact of prices as well as geographical and technical differences. In addition, the efficiency of the material input can be relevant (e.g. scrap or wasteful use of resources). Thus the same analysis as for personnel costs should be conducted for the material.

4.2.1.2 Functional Benchmarking

Step 2
In step two of the cost analysis, the profitability gap is not analyzed in comparison with an entire (or hybrid) company. Instead, individual functional units are investigated to confirm the optimization potential derived in step one and to define a productivity enhancement program. Departments (e.g. R&D, IT, Procurement, Sales) typically are a suitable proxy for functions and processes. It is important that the examined functional units are clearly defined in terms of personnel and cost volume and that they are mutually exclusive and collectively exhaustive. This is the only way to avoid that some functional areas are covered twice and others not at all. In practice, a cost center delineation has proven useful, even in the case of process-driven organizations.

For illustrative purposes it is again useful to compare the company with the leading competitor, or, ideally, with a hybrid company composed of all the best-in-class functions (figure 4-10). The same caution pointed out in steps one and two should be used here, as well: standardization of functional areas may lead to distorted results and the possibility that some facts are not represented in a one-to-one relation.

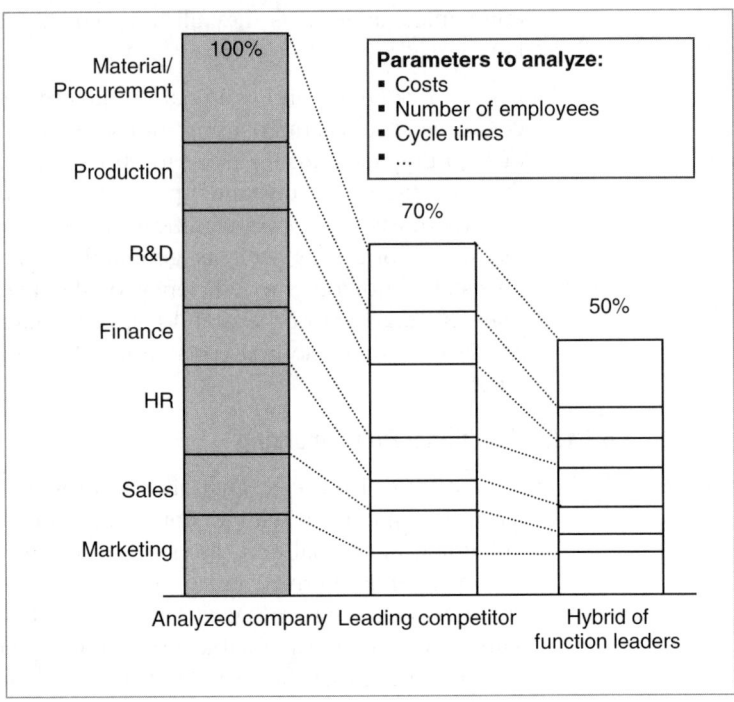

Figure 4-10:   Optimization potentials resulting from functional benchmarking

Determination of Volume Drivers   Functional benchmarks require the calculation of volume drivers to support the quantitative statements regarding costs. Product development time, for instance, is an important factor for R&D departments, sometimes even as significant as the cost targets. Other examples of appropriate benchmarking metrics include the number of employees served by an HR-specialist, the cost per million dollars of procurement volume, the number of operations the accounting department performs per cost pool, etc. This breakdown can be continued to any arbitrary level of detail. Manufacturing, for example, might measure production overhead or compare the efficiency of material used per employee.

Performance indicators and metrics do not need to be as highly detailed as possible in order to assure a meaningful functional benchmarking – the higher the level of detail, the more difficult it is to obtain comparable figures across companies. Rather, it is more important to select meaningful metrics and average them appropriately, if necessary. At the end of the day, the goal is to identify the causes for cost differences in order to derive opportunities for improvement.

The following example, engaging material costs, illustrates how a functional benchmarking is conducted and how improvement potentials are derived. Enormous potential for cost savings often exists in the area of procurement. In principle, cost savings can be achieved in two ways (figure 4-11):

	Price Reduction	Design-to-Cost
Starting Point	Commodity	Assembly groups
Time Frame	Short- to medium-term	Medium- to long-term
Potential	Low to medium	High
Measures	• Supplier concentration • Bundling • Renegotiation of contracts • Globalization of supplier structure	• Complexity reduction • Merging of modules • Product design

Figure 4-11   Starting points for a reduction of material costs

1. **Price reduction** aims at realizing the price reduction potentials at the supplier end. Ways to achieve this can be a concentration of suppliers, the introduction of eProcurement systems, or the geographical expansion of the supplier base to include those in low wage countries. These measures usually take effect in the short- to medium-term but are typically limited to commodity products and stan-

dardized parts. The cost savings potential depends on the maturity of the market and the position in the supply chain. Based on experience, the potential here ranks from low to medium.

2. **Design-to-cost** summarizes measures that aim at reducing the volume usage and the complexity of material and components. This includes product design optimizations, ways to reduce assembly times and costs, and steps to standardize specifications. Design-to-cost measures affect the areas of material and manufacturing. They do sometimes involve long lead times and high up-front investments, but the cost reduction potential is often significantly higher – again depending on the extent of value added.

The cost optimization potential in the area of materials can be estimated in three steps. The heads of Purchasing, R&D, and Manufacturing should be closely involved. Figure 4-12 illustrates these steps:

1. PRODUCT CLASSIFICATION

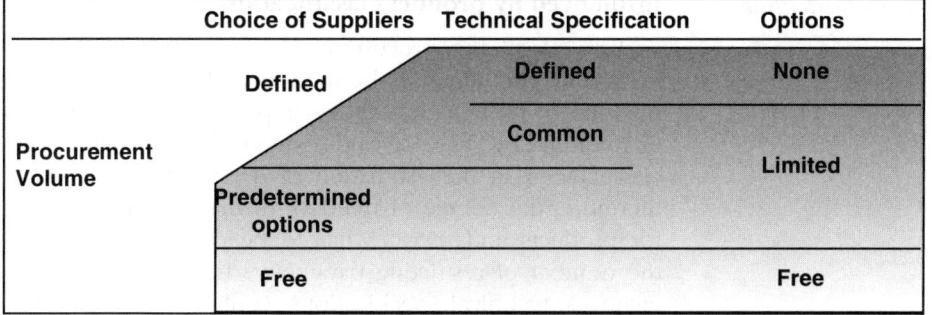

	Choice of Suppliers	Technical Specification	Options
Procurement Volume	Defined	Defined	None
		Common	Limited
	Predetermined options		
	Free		Free

2. ASSUMPTIONS BASED ON EXPERIENCE

	Price Reduction[1]	Design-to-Cost
Consumer Goods	13%	0%
Semi-Finished Prod.	7%	0%
Components	8%	8%
Assembly Modules	9%	16%

3. ESTIMATION OF POTENTIAL

	Share of Total Material Costs[2]	Influenceable Share	Potential on Influenceable Volume[3]	Potential on Total Volume
Consumer Goods	17%	80%	13%	1.8%
Semi-Finished Prod.	22%	65%	7%	1.0%
Components	37%	75%	16%	4.4%
Assembly Modules	24%	35%	25%	2.1%

X ... X ... =

Potential for Cost Reduction on Material Costs:	9.3%

[1] Figures based on Accenture experience
[2] Values of GreatValue, Inc.
[3] Figures based on Accenture experience (sum of price reduction and design-to-cost)

Figure 4-12: Exemplary estimate of cost reduction potential of material costs

1. **Determination of the procurement volume that can be influenced by product classification**

First of all the procurement volume (products and services) that can generally be influenced by optimization measures has to be determined. It has proven useful to categorize this volume by stage of value added, as it constitutes the decisive factor in the estimation of the savings potential. Furthermore, the degree of freedom of the procurement options for products and services has to be determined: The lower the number of restricting parameters (e.g., number of manufacturers, technical specifications, contractual obligations) the more flexible a company is in terms of externally procured products and services. If, for instance, a public tender asks only for a computer, the company has a wider range of supplier choices than if there are specific requirements in terms of processor speed, hard-drive capacity etc.

2. **Estimation of savings levers based on experience**

According to the possible approaches for reducing material costs discussed above (figure 4-11), there are two possible measures for cost reduction in the area of products and services that are procured either freely or subject to some restrictions. One is pure price reduction, the other design-to-cost. Experience-based assumptions about the savings potential that can typically be realized at each level of value added are assigned to the different measures. Figure 4-12 lists such empirical values, based on Accenture experience, for the four procurement categories of GreatValue. Subsequently, detailed savings potentials can be determined. The absolute savings potential increases with the complexity of the supply chain position, because the savings of the previous stages accumulate. Moreover, with increasing depth of value added, the savings potentials of design-to-cost measures grow faster than those of price reduction measures.

3. **Combination of influenceable volume and levers for savings potential estimation**

The share of total material costs corresponds to the procurement volume categorized by the degree of value added. Multiplying this share with the influenceable volume (refer to step 1) and the sum of the realizable cost savings per procurement group (step 2) results in the absolute savings potential per stage of value added. The sum of these represents the total potential for material cost savings. Depending on the composition of the procurement volume, this potential typically amounts to between 5 and 15 percent of the total material costs.

4.2.1.3 Detailed Drill-down Analyses

Step 3

The detailed drill-down analyses validate the identified profitability gap by offering detailed factual evidence in exemplary areas. Because of the often tight schedule of such projects, the results of functional benchmarks cannot usually be solidified to unassailable statements that are unanimously accepted by all departments and functions. Rather, these results more closely resemble and estimate according to the 80/20 rule (80 percent of the effect with 20 percent of the effort). Therefore, exemplary (and, ideally, representative) detailed drill-down analyses are used to substantiate the main results and statements. This way, the estimated top-down potential can be validated even on the most detailed level, at least in selected areas.

The functional benchmarking of a company in the aerospace industry revealed that the competitor's procurement costs were up to 15 percent lower. A comprehensive analysis of the entire procurement processes and material groups was not possible at that point because of the enormous complexity and time pressure. With the help of a public tender for an individual consumption item, it could be shown that significant cost reductions were possible by breaking the existing supplier monopoly. For example, hosts of Internet auctions typically state that 20 percent savings potential are possible through supplier base diversification.

The restructuring of a shipyard yielded another good example of a detailed drill-down analysis: traditionally, procurement only purchased red electric cables. Ships on the other hand were delivered exclusively painted in white. Because there was no feedback between manufacturing and procurement, the company incurred the costs of having to paint the cables in white. On top of that, red cables were actually more expensive than white ones. The cost reduction potential of improving the procurement management amounted to up to US$150,000 per ship.

These examples demonstrate that improvements are possible even in areas and departments that are already considered "optimally" managed – at least from the point of view of individual department heads. Thus, such analyses validate the estimated potentials and point to further areas of improvement.

4.2.2 Design of Program Components

Phase 2

After identification and validation of the profitability gap in phase one, the concrete optimization potential and realistic target costs for each cost pool have to be determined. For each cost pool, starting points and approaches for the cost reduction initiatives have to be identified. This process takes place in four steps (figure 4-13):

Step 1

Insufficient productivity is often coupled with over-capacity in many companies, especially in a weak economic environment. In order determine the real cost reduction potential, the current cost structure first has to be adjusted for any excess capacities (figure 4-13, step 1). As a first approximation, excess capacities of 20 percent would thus result in a reduction of the variable cost base by 20 percent. When eliminating over-capacities, how-

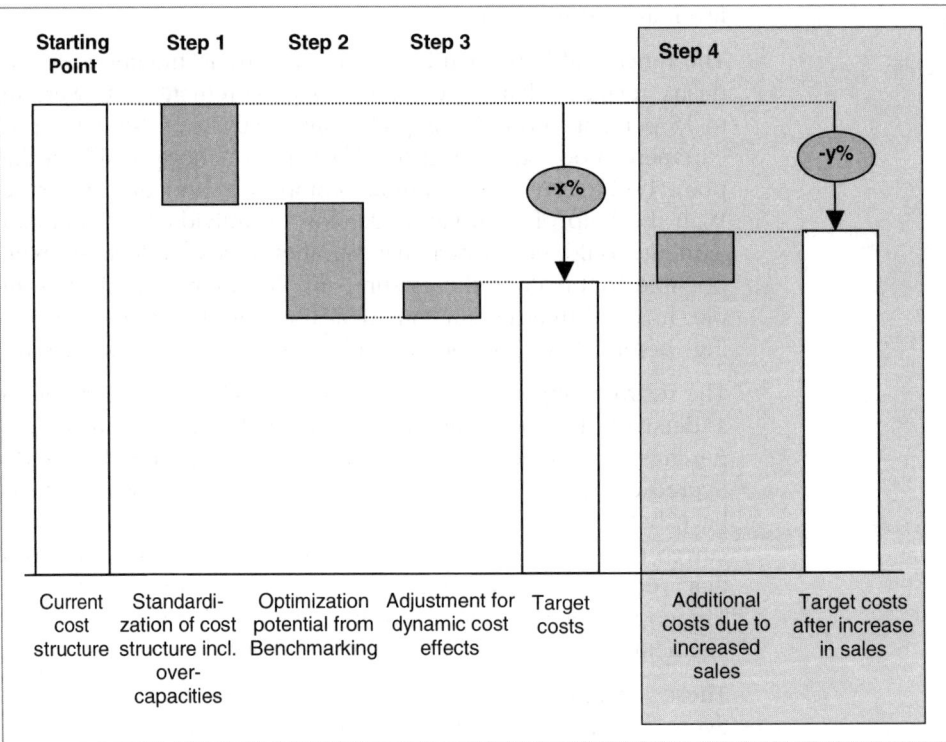

Figure 4-13: Calculation of the optimization potential

ever, relevant effects such as economies of scale or semi-fixed costs have to be taken into account, which may decrease in significance. The adjustment of over-capacities is necessary; otherwise the improvement program would be based on too low an optimization potential.

Step 2

In a second step, the results of the benchmarking analyses are applied to the adjusted cost structure (figure 4-13, step 2). It is often useful to validate the data by generating a zero-based cost model. This method involves constructing a hypothetical company with the same performance parameters from scratch. This construct then serves as a model with an optimized cost base at equal market performance.

Step 3

Analyses of cost structures are always static and drawn from historical data. Thus, step three amends the cost structure by including dynamic effects (figure 4-13, step 3). Such an adjustment – for example, caused by a drop in material prices – can reduce total costs, or increase them, for instance, in the case of wage raises. Figure 4-13 depicts the typical case of a net increase of the overall cost structure caused by increases in factor costs.

Step 4

The fourth step then adjusts the cost basis to compensate for changes in resource requirements due to altered market forecasts (figure 4-13, step 4). Here, too, a zero-based cost model is useful for comparison purposes as it takes economies of scale and semi-fixed cost effects due to changes in resources required into account. Especially in case of favorable market prognoses, it is advisable to use conservative growth figures in calculations.

Once the target costs are determined for all cost pools the program components have to be designed. They have to include precise starting-points and targeted approaches for the optimization. The action plans agreed upon at this point not only have to be implemented globally, i.e. company-wide, but also have to be detailed for all business units and functional areas, which in turn must participate in developing the plans. The result of this phase are individual project outlines for each business unit or functional area. These outlines include the targeted optimization potentials as well as approaches for their realization. For Manufacturing this could be "reduction of manufacturing cycle times by 30 percent," or for Material and Procurement this could be "material cost reduction of 20 percent via design-to-cost."

4.2.3 Development and Adoption of Detailed Measures

Phase 3

Based on the program components, a holistic profitability improvement program is now designed and approved (figure 4-14). Optimization targets are mapped out for all cost pools or functional areas so that they are mutually exclusive and collectively exhaustive. The measures to be taken within each program component can consist of technical modifications as well as process changes.

For each program component, cost reduction targets are set, a responsible leader is appointed and a time frame with milestones is worked out. The target potentials of a measure, even if it is stated in terms of non-monetary parameters, must be quantified (e.g., reduction of the workforce, shortening of throughput time, etc). The individual activities for each program component should be recorded exactly on a standardized form. These forms serve as specific how-to guides for all employees involved. Potentials can only be realized if each employee knows precisely what to do (differently) in the future and how to do it. A detailed description of exemplary measures and actions is given in chapter 6.2.

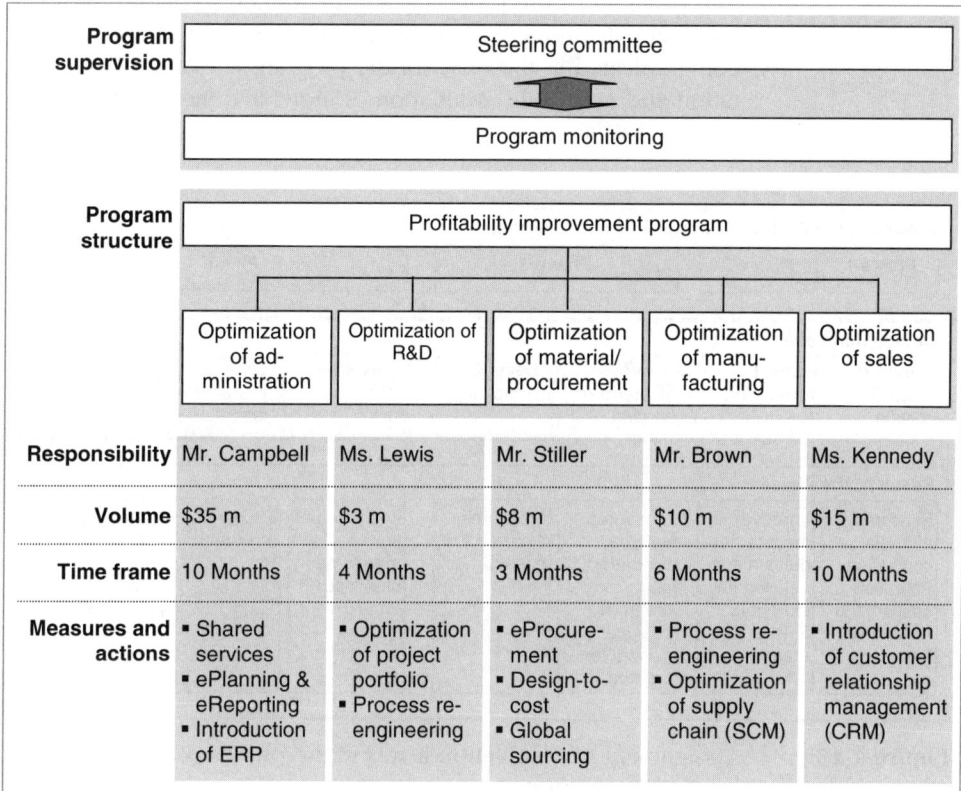

Program supervision	Steering committee				
	Program monitoring				
Program structure	Profitability improvement program				
	Optimization of administration	Optimization of R&D	Optimization of material/ procurement	Optimization of manufacturing	Optimization of sales
Responsibility	Mr. Campbell	Ms. Lewis	Mr. Stiller	Mr. Brown	Ms. Kennedy
Volume	$35 m	$3 m	$8 m	$10 m	$15 m
Time frame	10 Months	4 Months	3 Months	6 Months	10 Months
Measures and actions	▪ Shared services ▪ ePlanning & eReporting ▪ Introduction of ERP	▪ Optimization of project portfolio ▪ Process re-engineering	▪ eProcurement ▪ Design-to-cost ▪ Global sourcing	▪ Process re-engineering ▪ Optimization of supply chain (SCM)	▪ Introduction of customer relationship management (CRM)

Figure 4-14:   Overview of a profitability improvement program

A program-wide steering committee has to be established on the board level that regularly supervises the progress of the program. The operational supervision (program monitoring) requires the establishment of an independent corporate function.

The set-up of a profitability improvement program is not successfully completed before all involved units and divisions have internalized the significance of the goals and approaches. The more exact and plausible the analyses of the first two phases, the higher the degree of support even in those areas that are deeply affected (e.g. by resource reductions). It is thus advisable to commit all business unit heads to the program goals by having them sign a program document.

4.2.4   Implementation and Monitoring

Phase 4

The progress of the profitability improvement program has to be continuously monitored. A stringent, standardized set of rules, communicated at the onset of the program, eases the implementation and aids in the evaluation of individual measures.

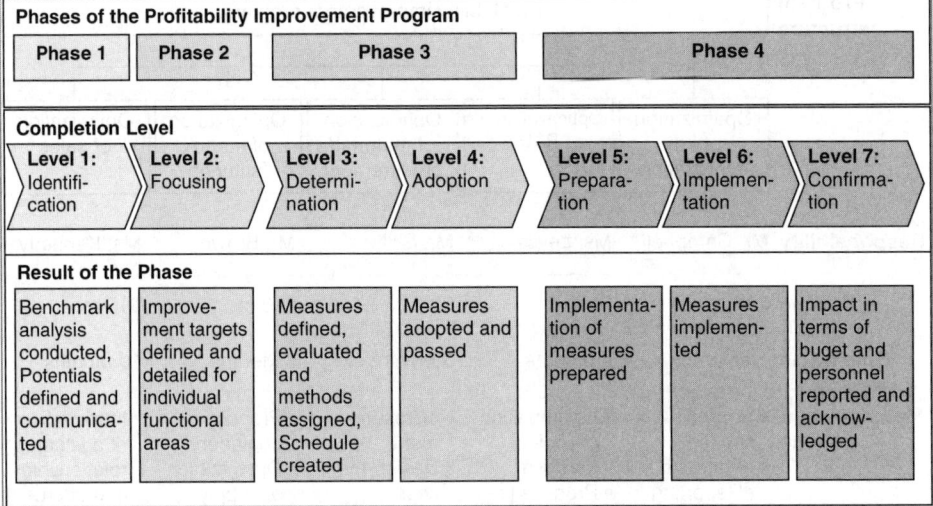

Figure 4-15:   Assignment of completion levels to the phases of the profitability improvement program

The person in charge of each measure records when a certain level of completion is reached. The formal status change, however, can only be entered by an independent entity such as the program monitoring or the steering committee. This separation is important in order to assure constant monitoring. Completion levels are the central elements that monitor the success of the four-phased process. As shown in figure 4-15, the seven completion levels can be assigned to the four program phases without ambiguity:

- **Completion level 1 as result of phase 1**
  Level 1 is reached when the benchmarking analysis has been conducted and the potentials have been identified and communicated.

- **Completion level 2 as result of phase 2**
  Level 2 indicates the successful completion of the second program phase. The profitability targets, detailed along the functional areas, have been set. The profitability improvement program and the corresponding sub-programs are all set up.

- **Completion levels 3 and 4 as result of phase 3**
  Level 3 shows that detailed measures have been identified and evaluated within the sub-programs. Level 4 is reached once these measures have been fully evaluated and documented and as soon as all participants have agreed on their feasibility. Furthermore, an implementation plan and timeline has been created at this point. Participating in the implementation are not just the managers of the business units directly affected, but also those responsible for the units to which tasks are outsourced or which receive services that are being optimized under the program.

- **Completion levels 5 through 7 as result of phase 4**
  The fourth program phase is divided into three areas: Level 5 indicates that the implementation requirements (for instance funds for acquiring new computers or software) are met so that the measure can be implemented immediately. At level 6, the measure has been implemented (e.g., the computer has been purchased, or the process changed). Level 7, finally, shows that the measure has been fully implemented and that cost savings are already in effect.

A program monitoring office is in charge of monitoring the progress of the entire program. For this purpose, an independent staff position should be established as early as in the third phase. Program monitoring is divided into measure monitoring and budget monitoring (figure 4-16).

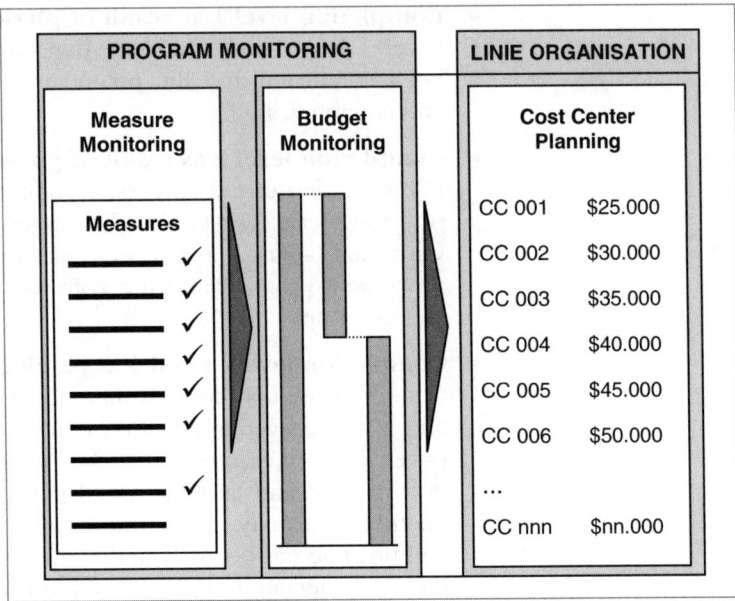

PROGRAM MONITORING		LINIE ORGANISATION
Measure Monitoring	**Budget Monitoring**	**Cost Center Planning**

Measures

CC 001	$25.000
CC 002	$30.000
CC 003	$35.000
CC 004	$40.000
CC 005	$45.000
CC 006	$50.000
...	
CC nnn	$nn.000

Figure 4-16:   Elements of the program monitoring

Measure Monitoring

Measure monitoring follows the progress of the entire program as well as of all individual measures. An important function is the supervision and adjustment of completion levels. With these the measure monitoring assesses the successful accomplishment of a measure. Arising problems are solved and communicated to the steering committee if necessary. Achieved savings are quantified and forwarded to budget monitoring.

Budget Monitoring

Budget monitoring is responsible for transferring the achieved savings onto the cost structure and allocating them to cost centers. Adjustments to meet planned sales growth also have to be considered here. The future budgets for the cost centers are planned on this basis. Because the budgeting process is dynamic, budgets might need to be updated several times. Budget monitoring transfers program results back to the operational responsibility of the line organization.

Through the optimization of its profitability and thus the achievement of operational excellence, a company creates the foundation for healthy growth.

4.3 Securing the Future Through Growth

Growth is not only a value creation lever. Macroeconomically, growth generates jobs. Microeconomically, it ensures a company's competitiveness. Dynamic growth is a significant success factor and strongly impacts the professional environment: success motivates, releases energies, and attracts a highly skilled workforce.

4.3.1 Managing Growth Horizons – Balance Ensures Sustainability

In order to ensure sustainable growth, the growth option pipeline – defined as the portfolio of increasingly concrete business ideas – must be continuously refilled. Here it is crucial to balance the individual time horizons.

4.3.1.1 Growth Option Pipeline and Horizons

The growth option pipeline defines the business portfolio along the growth horizons. The future product portfolio of a corporation is directly derived from it. Revenue growth is realized along three growth horizons (figure 4-17):

Horizon 1

Horizon 1 consists of the core business and therefore the revenue drivers that currently determine corporate profits. Businesses within this horizon can be pictured as fruits of grown trees that can be (sales-wise) harvested. It is quite possible that these businesses still harbor some growth potential; in the mid- and long-term however, new business ideas need to be identified and realized in order to consistently fulfill growth expectations.

Horizon 2

Horizon 2 encompasses potential businesses that do not contribute significantly to today's business, but some of which will support the company's results in the mid-term. They correspond to sprouts in the growing stage.

Horizon 3

Horizon 3, finally, contains ideas and growth opportunities of which some will be implemented in the future and thus will significantly contribute to company profits; numerous others, however, will never be realized – as with the multitude of seeds that will only produce a few grown trees.

The time frame of these horizons varies widely from industry to industry. While the half-life of current revenue drivers (first horizon) in traditional sectors can be measured in years, in more dynamic industries the half-lives frequently only represent months. As illustrated in figure 4-17, the growth horizons are not distinctly separated from each other, but should rather be perceived as a cascade.

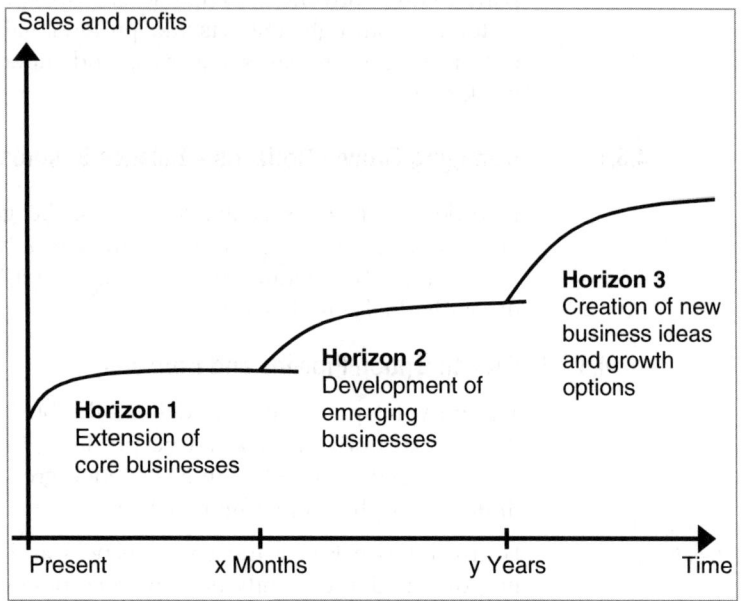

Figure 4-17:   Structuring the growth pipeline over horizons

In order to secure consistent growth, all three horizons have to be constantly and actively managed. This means that a company must simultaneously protect current profit drivers and systematically nurture potential (new) businesses – to secure its future options. The company leadership is always faced with the challenge of shaping and managing the three horizons in a balanced manner.

Value Leaders
Utilize Synergies Successful companies (value leaders) excel by utilizing the synergies that result form a balance of the different horizons: established companies with well-developed, profitable businesses possess a solid foundation in the first horizon. This sustains research and development for new businesses and it solidifies the corporate reliability in the perception of investors. Young companies create strategic degrees of freedom in the second and third horizon by developing creative employees and fueling the fantasies of capital markets. Based on the different positioning,

established companies often lack growth opportunities, while emerging companies often do not have a strong foundation.

How can one recognize a balanced growth portfolio? An analysis should follow along the following key questions:

Horizon 1

The current revenue drivers are mainly measured against the competitors.

Key Questions Horizon 1

➢ Do the current revenue drivers generate sufficient profits in order to enable investing in future growth?

➢ Are the metrics and controlling instruments performance-oriented?

➢ Is the market share at least stable?

Horizon 2

The potential businesses in the second horizon are measured according to whether they can grow to become revenue drivers in the foreseeable future.

Key Questions Horizon 2

➢ Will the investment in this sector generate returns in the foreseeable future?

➢ Do these sectors attract enough new employees so that the growth can be realized?

➢ Do the investors place trust in these new businesses?

Horizon 3

Securing a sufficient number of future business ideas must also not be neglected.

Key Questions Horizon 3

➢ Does the management commit enough time for the development of new business ideas?

➢ Are innovations really new businesses or merely extensions of existing products and services?

➢ Are concrete steps being taken to consistently transform new ideas into business options?

The correct balance of the growth portfolio is largely determined by external factors. The dynamics and evolutionary pace of the sector, industry risk and volatility all play a significant role.

Figure 4-18 illustrates the necessity of constantly refilling the growth option pipeline. The growth of revenue drivers within the current business portfolio will stagnate in the third horizon at the latest. In the example shown this would take place in about two years. Therefore in order to attain long-term growth or even to increase the growth rate, potential future businesses have to be developed and new business ideas have to be conceived already today. Not all of the ideas will develop into potential businesses. Experience shows that only a fraction of them, a maximum of one out of eight ideas, will become a revenue driver. In order to attain a revenue growth of e.g., US$100 million in the long run, the growth option pipeline already has to be filled with business ideas with a potential value of US$800 million, at a minimum.

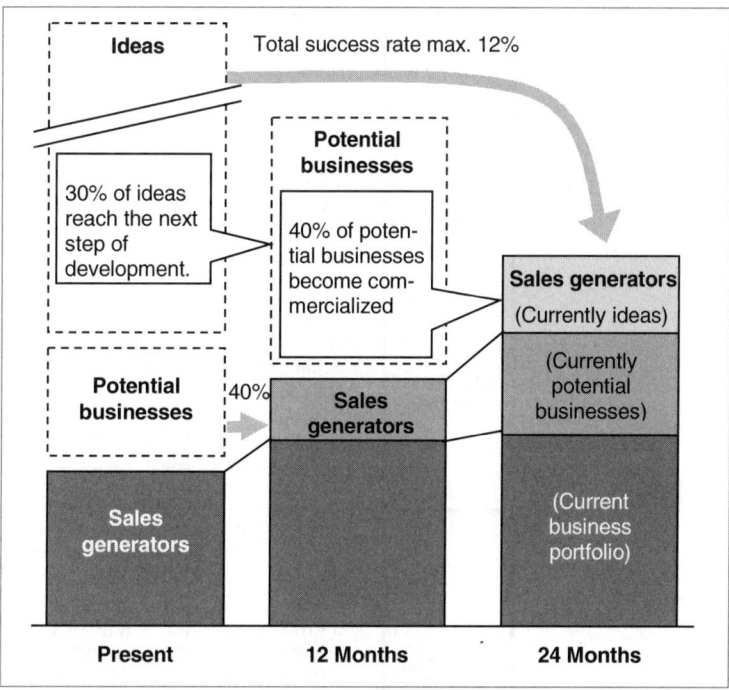

Figure 4-18:   Requirements for the business pipeline to ensure an adequate business portfolio in the long-term

All three growth horizons necessitate different management principles, management styles, planning approaches, and metrics. This is shown in figure 4-19.

Horizon 1 emphasizes profit maximization from existing businesses. This necessitates a high measure of experience on the operating side. In this instance, success is primarily gauged by efficiency and profit. For the development of emerging businesses in Horizon 2, entrepreneurial spirit and strategic skills are primarily required. The primary objective and thus the key performance indicator in this case is growth. The creation and development of new business ideas and growth options in Horizon 3 requires creativity, unconventionality and innovative spirit. Numerous business initiatives with an expected value of profit are the most significant success measures in this case.

	Horizon 1	Horizon 2	Horizon 3
Leadership style	• Low tolerance for errors • Experience in operative business	• Decisive • Entrepreneurial • Skilled in dealing with volatility and change	• Unconventional • Visionary • Assertive
Management principles	• Fulfillment of plan targets • "Milking" of existing product line-up	• Support for creativity and autonomy – "long leash"	• Strengthen visionary thinking and conduct – "think big"
Planning methods	• Strategies for retention of current market share and profitability improvement	• Business development strategies • Business plans	• Scenario analyses • Project plans and milestones • Gradual build-up, development and financing

Metrics and Incentives	• Incentives for short-term success
	• Metrics:
	  o Productivity
	  o Efficiency
	  o Profit margin
		• Incentives for creating growth
		• Metrics:
		  o Revenue
		  o Market share growth
		  o Client acquisitions
			• Career opportunities and responsibility for newly developed business sector as incentives
			• Metrics:
			  o Number of initiatives
			  o Success probability
			  o Expected value of profit

Figure 4-19: Different requirements of the three growth horizons

4.3.1.2 Growth Options

There are many ways to grow. The different approaches to growth can be divided into three categories:

Organic Growth

This category encompasses all the activities through which a company can expand on its own accord. Examples include the addressing of new customer groups or the entering of new geographic markets, the development of alternative distribution channels, or even of new products and services. Organic growth alone – dependent on market environment – can typically generate growth rates of five to ten percent annually, and of course significantly more in young industries.

Among high-tech companies, the market leader Intel provides an example for value creation through intensive organic growth. During the 80s growth was stimulated by applying aggressive sales strategies. One of these marketing offensives, the so-called "Operation Crunch", persuaded IBM to use the Intel 8086 processor in its products, instead of the Motorola 68000 processor that, in the opinion of industry experts, was technologically at least as good as the 8086. In addition, Intel was the first CPU manufacturer to market directly to the consumer, striving for product differentiation with its "Intel Inside" campaign. Thus, with Operation Crunch und Intel Inside, the corporation laid the foundation for its dominant role as component supplier for PCs.

Moreover, the parallel development processes, in which several processor generations are going through different stages of the R&D pipeline at the same time, enable Intel almost always to be first to introduce the newest processor generation to the market. Furthermore, Intel now leverages its brand recognition and market power aggressively to move into the quickly growing market for PC add-ons (PC cameras, audio players, etc). With this strategy, the market leader in main processors is also strengthening its core business: "The value of the PC increases exponentially as we have more functionality attached to it" (Craig Barrett, CEO Intel).

Particularly the value leaders do not always transition smoothly between the growth horizons, but sometimes undertake significant restructuring efforts. For instance Intel's departure from the DRAM market in the mid-80s amounted to nothing less but a complete redirecting of the entire company.

Similarly the Finnish company Nokia is a good example of growth through excellent management of the product pipeline. Always on the lookout for innovative technologies and therefore new growth markets, the former rubber producer has undergone some astounding change since 1865. Nokia has been evolving through rubber, chemicals, cables, electric power and electronics products to become a leading solutions provider for wireless communications.

Venturing

The newest – and perhaps the most exciting path to growth – is venturing. This approach is based on such instruments as incubators or venture capital, either externally or internally, e.g. in case of corporate venturing. This venue represents a longer-term approach to company growth and primarily creates growth opportunities in the second and third horizon.

For quite some time now, companies like Cisco Systems have been investing venture capital in up-and-coming, promising start-ups to secure quasi "purchasing options." Internal commercializing is also a part of venturing: Bell Laboratories, for example, is an incubator for new technologies on which its corporate parent, Lucent Technologies, has relied on frequently.

External Growth

The capability to acquire highly innovative companies and integrate them within the shortest timeframe is one of the basic prerequisites for successful growth. But also partnerships and strategic alliances are constantly gaining significance in numerous industries. At the center of these activities is not the acquisition of physical assets but rather of market-ready products, commercializeable ideas and highly qualified personnel. Theoretically, there are no limits for external growth, but in practice it is extremely difficult for any organization to successfully sustain growth rates of more than 50 percent annually over the long-term.

The network supplier Cisco Systems is a frequently quoted example of a growth company that relies on its aggressive acquisition strategy. Since 1993 Cisco has acquired over 70 companies in transactions worth more than US$35 billion and thus gained almost 6,000 highly qualified employees. The objective was not only to maintain and expand its technological leadership in core businesses, but primarily the systematic development of new markets. Based on its dominant position in a data networking sub-market (LAN-, WAN-routers), Cisco expanded through acquisitions into the sectors of: network management, security solutions, network-access systems, and fiber optic networks. Recently the focus was primarily on companies involved in Internet protocol-based voice transfer (voice over IP – VoIP). In sum, Cisco has developed a dominant position in the converging world of telecommunications- and data network by acquiring businesses like Sentient Networks, Komodo Technology and Calista, thus capturing significant future growth potential.

In addition to the acquisition activities, Cisco is very actively involved in building its "New World Ecosystem," a worldwide network of alliances and partnerships. The purpose of such alliances is twofold: on the one hand, to capture new markets – as in the case of the "Invisix" alliance with Motorola to develop GPRS and UMTS cellular communications networks. On the other hand, Cisco aims to integrate those sectors of the value chain, which the company does not cover itself – e.g. the Operation and Business Support Systems (OSS/BSS systems) or network integration. This approach not only enables Cisco to offer one-stop-shopping solutions but also to influence industry standards.

The following chapter provides an overview of options for filling the growth option pipeline and thus ensuring sustainable growth.

4.3.2   Venues to Growth

Investments and focused innovations are basic requirements of growth. An insufficient investment volume or the distribution of investments over too many technology sectors can provide the explanation for growth gaps. The art is to secure growth options in numerous promising sectors, thus keeping a foot in the door, while avoiding a shotgun approach and thus below-critical mass in significant, investment-intensive areas. Here, too it is helpful to conduct a benchmarking of investment parameters against key competitors.

In order to benchmark a company's investment activity against those of competitors, it is advisable to allocate the total investment volume to individual business units or technology sectors in a roughly proportional fashion. In this context the term "technological sector" can be defined with varying scope. If, for instance, one compares the large telecommunications and network suppliers, a relatively rough division into optical networks or third-generation cellular communications could be fully sufficient. However, if one analyzes smaller or more specialized companies or their individual business units, then the scope will have to be much more narrowly defined.

Figure 4-20 makes it quite evident that GreatValue, Inc. distributes too small of an investment volume onto too many sectors. This renders the average investment rate per technological sector insufficient.

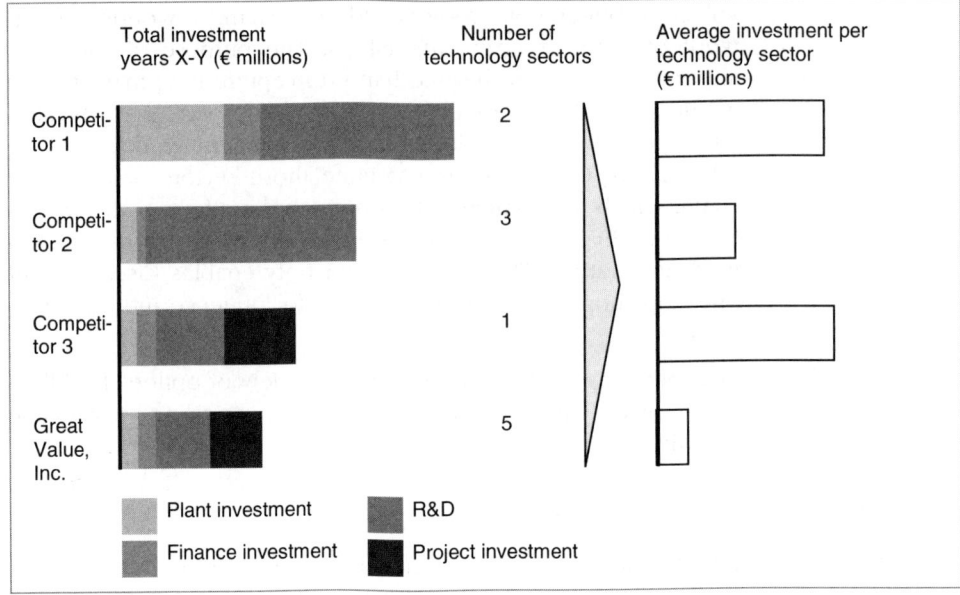

Figure 4-20: Investment activity at GreatValue, Inc. compared to main competitors

The next step consists of dividing the investment activity according to the three venues to growth: there are numerous approaches within organic growth, venturing, and external growth a selection of which will be described in detail below.

4.3.2.1 Organic Growth

Increasing revenue from existing customers on established products is the most obvious growth dimension. Thus, for example the modification of a unique selling proposition can lead to an increase in revenues. Other approaches may engage marketing measures or customer loyalty programs. The Internet allows not only individualized, targeted advertising, but also more granular customer segmentation. In addition, establishing online communities can strengthen customer relationships. E-Marketing and Customer Relationship Management (CRM) are established methods in this area. The Internet bookstore Amazon has been successfully employing the concept of community-based selling since 1997. This concept has been patented for online book selling and, according to the Financial Times, accounts for about

30% of Amazon's sales. Because of lower maintenance costs, this target-group specific business segment is also significantly more profitable than the others.

Once existing customers are sufficiently loyal to a company, the second stage of organic growth could entail the extension of the customer base. Here too, the Internet is of increasing significance in reaching new customer groups or advertising more aggressively. The geographical expansion of business activities and the extension or bundling of new distribution channels is another growth option based on an already existing service portfolio. Cross marketing, for example, allows advertising of product (or services) bundles from different complementary sectors to the same target group.

Through the Internet, these growth paths may be treaded best by integrating the company in so-called (online) communities. A community is an interest group, comparable to a fan organization or sports club in real life. Participation in these virtually structured interest groups means that companies can also gain access to participating target groups, to whom they can advertise their products and services and thus capture growth potentials. Online marketplaces also bring together groups with the same interests, and they connect manufacturers to buyers never thought of before. Many leading companies – 3M, Boeing, Dow Chemical, General Mills and Sharp, among others – for example, use the Internet marketplace yet2.com for offering and licensing licensable technologies. This not only extends their services to new customers, but also avoids costly R&D processes, increases speed-to-market, and maximizes R&D profitability.

An essential element within the core capabilities for generating organic growth is the capacity to develop new products and services. A successful growth program has to exhaust the improvement options in terms of new product introduction to the fullest extent. Not only have development time and time-to-market to be optimized. A comprehensive program must also incorporate strategic options such as platform strategies and parallelizing potentials. Chapter 6.2 presents a detailed catalog of a selection of such measures.

4.3.2.2 Growth Through Venturing

Venturing is a relatively new approach and it represents an interface between organic and external growth. Internal venturing fills the growth option pipeline through focused development, isolation, and commercialization of existing business ideas within a company. One instrument of the internal venturing approach is the business plan competition. Here, business and/or product ideas are identified and employees from various departments are involved in entrepreneurial thinking. This is particularly useful for employees in the research and development department, who learn to view a product not just from a technological perspective, but also from the vantage point of its market potential. Ideas collected in this manner can be commercialized within the framework of an incubator program. The Minnesota Mining & Manufacturing (3M) company has used this instrument to integrate the access for all employees to financing of their business ideas into the corporate culture.

A business plan is an indispensable tool for the evaluation of an idea, since it describes the strategy and the business case as well as the market environment, and it makes assumptions transparent. It enables third parties to evaluate an idea relatively quickly on a realistic basis. Box 4 details the basic elements of a business plan and presents an example from daily practice.

The internal potential of ideas discovered in the course of business plan competitions needs to be fed into a development pipeline in which the new business ideas undergo a process of staged funding. The financing volume is to be increased gradually as successive milestones are completed and targeted goals met. Thus hopeless endeavors are quickly filtered out with minimal losses and promising options are commercialized as quickly as possible. Ideally, an incubator with an external presence assumes these selection and control functions.

The venturing approach can also be externalized by setting up venture capital funds. There are numerous examples for this procedure, particularly among global players. According to venture capitalist Harry Edelson, quoted on the investor portal multex.com, companies have increased their venture capital funding from US$1.0 million in 1997 to US$16.5 million per year in 2000.

Box 4: The Business Plan

To some degree, the content of a business plan is tailored to the particular audience but usually contains the following elements:

- Summary,

- Description of the product or service, especially of its unique selling proposition,

- Introduction of the (project) team and its roles,

- Market and competitive analysis,

- Marketing and sales strategy,

- Organization and business architecture,

- Financial prognoses, and

- Opportunities and threats.

A good business plan analyzes the market environment, success factors, and forces that characterize a particular industry. In addition, it answers certain key questions: What are our company's capabilities and intentions versus those of our competitors? How will competitors react? How do our success factors compare with those of the competition?

Venturing, however, is not an unproblematic activity and should therefore only constitute an addition to the regular investment activities. Funds urgently needed for research and development or even for day-to-day operations must not be misappropriated for this purpose. Often, companies lack the skills and experience necessary to conduct rapid but thorough due diligence procedures or even business plan checks in the very volatile venture capital market. Thus, Edelson currently observes large industrial companies withdrawing from this sector as a consequence of the bitter losses they suffered on a broad front.

4.3.2.3 External Growth

As the example of Cisco Systems shows, external growth, especially facilitated through acquisitions, can contribute significantly to filling the growth option pipeline. In some cases, taking this path might even be unavoidable in order to ensure sustained growth. But how can a company that is stuck in a profitable, but stagnating industry identify new, promising growth businesses and suitable acquisition targets? Often, a substantial capital outlay is required for such an endeavor, which precludes any half-baked experiments. Rather, a systematic process for the identification and evaluation of potentially interesting business areas and acquisition targets is of the utmost significance. Such a process could be structured into five steps (figure 4-21).

Step 1

Initial Filtering of Feasible Business Areas

First prospective sectors/technologies are described as extensively as possible and evaluated according to predefined conditions and criteria. Here, the solution space or search scope can be very broad initially, in order to avoid missing a feasible option. Three dimensions span the search scope in the following example:

1. **Dimension: Industry Structures**
   The first dimension systematizes all existing industries. Among the sources for the collection of industries are market research institutes, leading research services, but also reports by investment banks and statistical offices.

2. **Dimension: Industry Value Chain**
   In the second dimension potential new areas of activity for the existing business units are charted along the value chains of the potential new industries. Sources for this activity include industry-specific publications as well as brainstorming workshops with the company's employees.

3. **Dimension: Key Technologies**
   In order to stress the importance of significant growth potential of a new business area sector, the third dimension charts out key technologies that are expected to grow explosively in the near future.

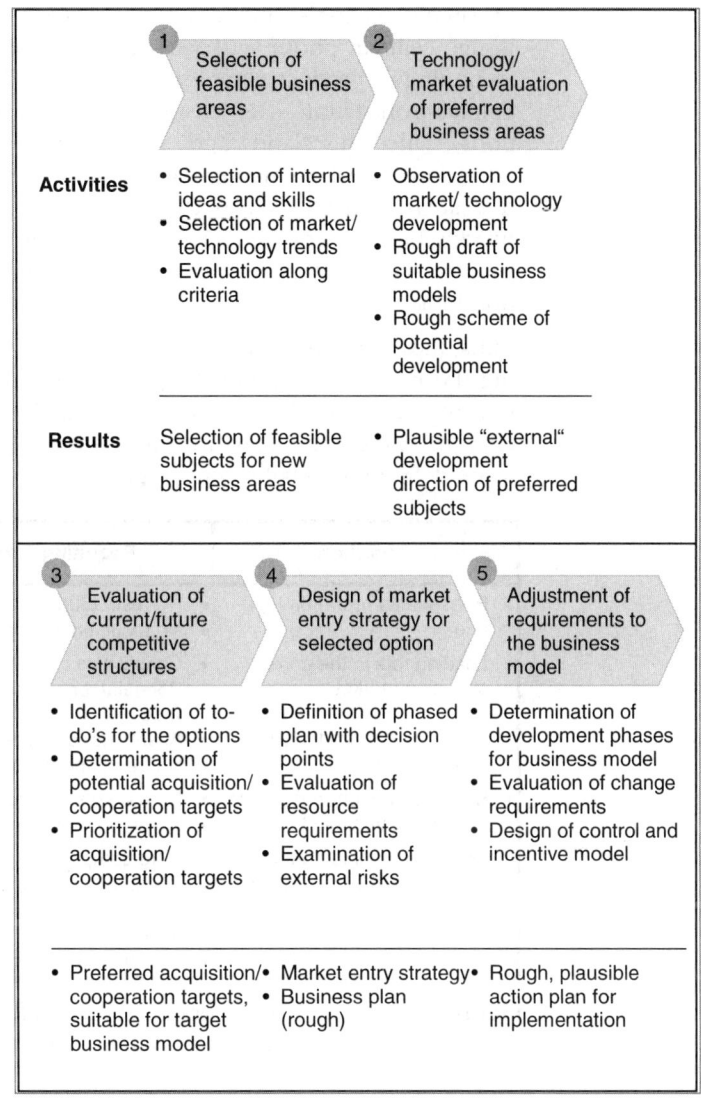

Activities	**1** Selection of feasible business areas	**2** Technology/ market evaluation of preferred business areas
Activities	• Selection of internal ideas and skills • Selection of market/ technology trends • Evaluation along criteria	• Observation of market/ technology development • Rough draft of suitable business models • Rough scheme of potential development
Results	Selection of feasible subjects for new business areas	• Plausible "external" development direction of preferred subjects

	3 Evaluation of current/future competitive structures	**4** Design of market entry strategy for selected option	**5** Adjustment of requirements to the business model
	• Identification of to-do's for the options • Determination of potential acquisition/ cooperation targets • Prioritization of acquisition/ cooperation targets	• Definition of phased plan with decision points • Evaluation of resource requirements • Examination of external risks	• Determination of development phases for business model • Evaluation of change requirements • Design of control and incentive model
	• Preferred acquisition/ cooperation targets, suitable for target business model	• Market entry strategy • Business plan (rough)	• Rough, plausible action plan for implementation

Figure 4-21:   Program for the identification of potential new business areas

It is helpful to visualize the product and service landscapes to illustrate all three dimensions. The various steps in the value chain allow the systematization of the range of a company's offerings. Here as well, the criteria are closeness to the consumer (a simple example: modem – dial-up server – Internet backbone; cellular phone – base station – switching and routing technology), product complexity, or demand for integration (example: workstation – server – network software – group applications). Once this "landscape" is filled with market data, growth rates and, possibly, market values of the key players, the attractive areas as well as convergence and substitution trends will quickly become apparent. This process gradually isolates feasible subjects for new business areas with the help of filters and iteration loops.

Thus, the spanned search scope is gradually narrowed down (using the necessary iteration loops, if required) to an initial selection of a few feasible subjects for potential new businesses areas. Of course, the parameters for this identification process first have to be defined in detail (figure 4-22).

Condition	Resulting Exclusion Criteria
• Reasonable return on revenue	• No restructuring scenarios • No focus on cyclical/commodity markets
• Leading role in the corresponding market	• Critical size in every sub-market • Potential for at least the #3 position worldwide
• Solid foundations for business activities	• Assets not easily recreated • No trend and "hot topic" subjects
• Build-up through reasonable one-time investment	• No markets dominated by large companies • No brand with high R&D initial investments
• Option of complementary and synergetic integration	• Core competency on similar foundations • Suitable culture and long-term goals • Suitable size (revenue US$250-750 million mid-term) • No consumer goods

Figure 4-22:　　Parameters for the identification of potential new business areas for Great-Value, Inc.

As soon as the business areas are staked out, a filtering process ensues. Potential areas are analyzed based on several exclusion criteria (filter criteria). The number of filters varies according to the situation. Figure 4-23 presents a possible example of a filtering process.

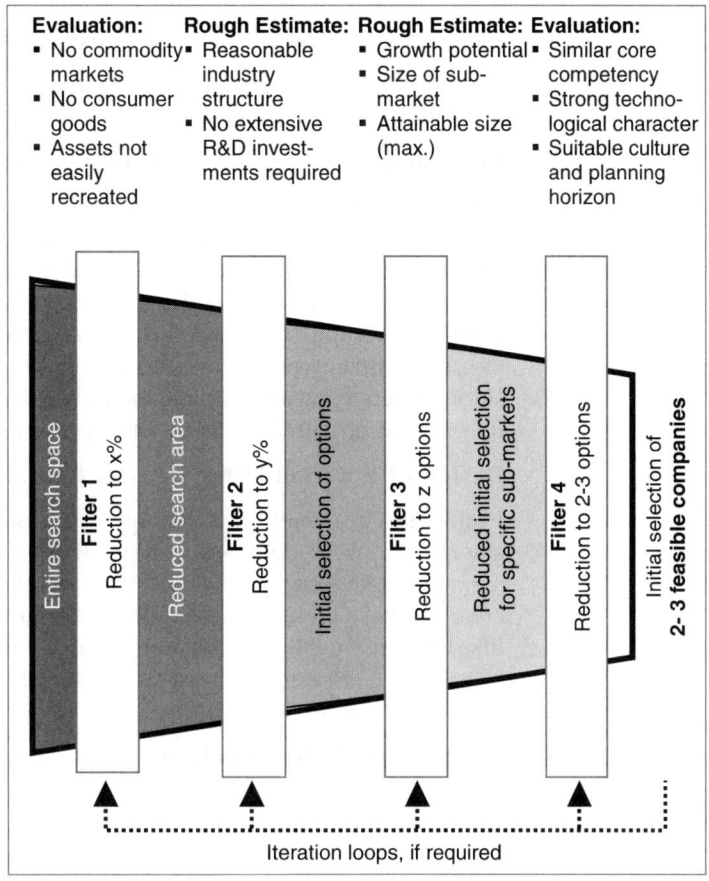

Figure 4-23: Filtering process to reduce the entire search space to an initial selection

Step 2

Technology and Market Evaluation for Identified Areas

In a second step, market-, technology-, and trend-research is conducted for the areas identified as promising. With the help of key questions industry value chains therefore have to be created and analyzed for every area:

Who are the typical providers in every segment of the value chain?

Who are the typical customers in each of these segments?

What are possible trends, for instance from a technological and regulatory perspective?

What business models are feasible?

What would be the optimal value chain depth?

The result of step 2 is the identification of a realistic development area for external growth.

Step 3

Evaluation of Current and Future Competitive Structures

This step is used to assess to what extent market entry through acquisitions is realistic. For all business models from the initial selection, the competitive structure has to be analyzed. For each competitor, the following data needs to be collected: revenue, number of employees, regional concentration, market capitalization or company value, and ownership structure. Realistic acquisition targets constitute an indispensable prerequisite for successful external growth. The result of step 3 is the list of such targets.

Step 4

Design of a Market Entry Strategy for the Selected Option

In this step a market entry strategy based on the identified acquisition options is developed. It is divided into several phases, aims to achieve sub-targets and is oriented toward reaching concrete results. The resource requirements have to be exactly quantified for every phase. The full extent of every business model has to be mapped using checklists. Finally the business plan is the product of this step.

Step 5

Adaptation of the Requirements for the Business Architecture

In this last step before the final decision, the corresponding requirements for the business architecture are evaluated and potential problem areas that are expected to arise during the integration phase are identified. All areas, which have been detailed in earlier chapters have to be taken into account within the business architecture, in particular: organization, processes, incentive systems, and human capital. Cisco Systems has been successful with its acquisition-based growth strategy, because it was capable of integrating acquired companies within the shortest timeframe - 100 days are frequently mentioned. This integration capability thus constitutes a core competence of Cisco.

Post-Merger Integration

The so-called *Post-Merger Integration* finally determines whether an acquisition strategy turns out success or failure. The company will incur high costs if the acquisition potential is wasted, because the synergies are not realized and acquired talent is not retained. According to a study by Prof. Dr. Schewe (University of

Muenster) and Dr. Johannes Gerds (Accenture) in 62 percent of researched corporate mergers the projected targets were not met. In most cases, senior management's lack of integration competency was responsible for the failure of mergers. Instead of carrying out a consistent integration management, many managers follow an array of "myths" that undermine successful integration. Box 5 confronts these myths with facts.

Box 5: Myths About Post-Merger Integration

Myth #1: "The faster the better!"

Wrong! – The faster the integration is carried out, the more doubtful are the prospects for success. High speed reduces the control over the process. Employees are overwhelmed and become demotivated. Necessary steps are carried out only partially or not at all. The risk of erroneous decisions increases. Possible advantages of a speedy integration, for example the utilization of change expectations or the reduction of insecurities of the employees cannot make up these disadvantages.

Myth #2: "Employee resistance creates the greatest obstacle to integration!"

Wrong! - Insufficient management skills represent the greatest obstacle to integration. Often, the local management is overwhelmed by integration-specific tasks. Many top managers fail particularly when it comes to interlinking the operating processes three to four months after closing the contract itself. Even though the lack of motivation on the employee side can strongly impede integration, its impact compared to missing management skills is relatively weak.

Myth #3: "Soft measures are more important than hard measures!"

Wrong! – The success of integration measures depends on context and objectives. There is no universal recipe that ensures success. Integration measures always have to be accompanied by caution. Poorly orchestrated application of measures can drastically reduce the chances for success. Especially an inappropriate selection and nomination of the new executive team can permanently damage the willingness of managers on both sides to cooperate.

Myth #4: "Mergers-of-equals are riskier!"

Wrong! – Mergers-of-equals are in principle not any more risky than mergers of companies with different strength. However, in a merger-of-equals the top management must actively support the integration process and provide additional managerial capacities for this purpose. Especially the inclusion of the employees in the integration process requires sufficient resources. Because of a thin resource blanket in many mid-sized companies these resources are frequently not even existent.

Myth #5: "Overhead synergies are easier to realize!"

Wrong! – More than a half of all mergers fail independently of the integration objectives. The management team who designs the integration process definitely also holds the key to its success.

In order to ensure the optimal employment of growth and profitability on a sustainable, long-term basis, the appropriate metrics, incentive systems, and control mechanisms must be installed at all the levels of the company. Therefore the following chapter describes how the pillars of the house are mounted on a solidly supporting floor of metrics and incentive systems.

5

Metrics and Incentive Systems:
The Supporting Floor

The earlier chapters have analyzed what significance profitability and growth have for a company and how the levers of the House of Value Creation can effectively increase shareholder value. The application of appropriate metrics and incentive systems supports, monitors, and evaluates the implementation of the company objectives on all levels. Therefore the relationships between the individual internal units need to be clearly defined in order to attribute the value contributions to the originators and to reward or sanction the individual and institutional achievements.

Even though the metrics represent the desirable state of affairs from the corporate standpoint, they are not sufficient to manage departments and employees. In addition, incentive systems need to be designed, which motivate employees to realize and reach these target metrics.

5.1 Operationalization of Corporate Objectives Through Concrete Metrics

Metrics

The basis for attaining the company objectives consists of increasing profitability and growth. However, individual employees or organizational units can hardly influence these two parameters. In fact, it is necessary to introduce sub-targets, which, when combined, make up the defined global profitability and growth targets. When defining the metrics, it is most important that each one reasonably measures a relevant management goal by itself. The complete set of metrics must represent the value creation levers comprehensively, but without redundancies or ambiguities.

Metrics should not be exclusively short-term oriented since this will interfere with the long-term corporate perspective. For example, in the course of a global cost reduction program, cutbacks in employee training programs or R&D expenditures rarely pay off in the long term. The declining learning curve would cause valuable innovation potential to be lost.

Metrics not only aid in the operationalization of corporate objectives, but also evaluate the performance of business units and employees.

Metrics represent both benchmarks as well as performance indicators. Based on these parameters, the company can evaluate, direct, and reward the performance of employees and organizational units on every level. They serve as standards for employees and investors alike.

When metrics and incentive systems that derive from them are defined, the individual company units must not be forced into a rigid corset of performance indicators. Instead, management has to aim at granting the individual units and employees as much entrepreneurial freedom as possible.

By the same token, not every unit should be governed by the same metrics. This is particularly true in conglomerates with heterogeneous divisions, since different industries frequently underlie different rules. Thus, the target metrics applicable to a specific industry depend to a significant degree on the corresponding industry clockspeed (Box 6).

Box 6: The Industry Clockspeed

The concept of an industry clockspeed describes the dynamics of an industry and in this context helps to correctly evaluate the business environment qualitatively as well as quantitatively. Every industry has its own life cycle.

The industry clockspeed is primarily shaped by the pace of an industry's technological evolution. To measure it, the clockspeed can be decomposed into its different dimensions, for instance, into the process-technological and product-technological component. The process-technological clockspeed is the time that typically passes until the manufacturing assets and processes become obsolete. The product-technological clockspeed could be measured according to the innovation cycle of a product, i.e., the interval between two product generations. But an industry is also impacted by other, non-technological clockspeeds. Examples include the speed with which organizational changes take place, distribution channels evolve, or brand names are created. Which clockspeed mainly characterizes an industry can vary sharply in every individual case.

In order to achieve a positive evaluation by the capital markets, a company must align every business unit with the applicable industry clockspeed, or, ideally, actively shape this clockspeed. Especially in the case of conglomerates, whose different business units might be subjected to different industry clockspeeds, this alignment is extremely complex. Here, the pace of the individual business units is usually not driven by the respective industry clockspeed, but by a uniform company clockspeed that is often dictated by the slowest industry the conglomerate is active in. This clockspeed is frequently too slow for units involved in more dynamic industries. In addition to lacking transparency, this is one of the primary reasons for the typical discount the capital markets assign to a conglomerate versus the sum of its components.

A successful company not only has to adapt to the industry clockspeed, but it can also actively shape it. This is possible especially in young and dynamic industries because of the high growth potential and the resulting opportunity to assume market leadership. The rapid industry clockspeed related to the high rate of innovation creates high market valuation potential for those companies who are perceived as capable of market leadership. Thus, if a company succeeds in convincingly communicating its industry-shaping role to potential investors, it usually obtains a premium valuation.

5.2 Principles of the Market Economy as a Basis for the Evaluation of Corporate Entities

In order to evaluate the performance of corporate entities, free-market principles need to be instilled into the entire organization. The organizational preconditions for an internal, market-oriented performance evaluation include:

- profit centers with clearly defined tasks, goals and a large degree of autonomy with the associated responsibilities, as well as

- fair transfer prices for products and services.

5.2.1 Profit Centers

Management and measurement are helped most if as many business units and other corporate entities as possible are organized as profit centers. Profit centers are units with responsibility for their own profits and losses. They therefore constitute "companies within a company" and can to a large extent be evaluated and rewarded competitively, i.e. according to outside market conditions. Even staff positions and service units, such as secretarial offices or cafeterias, can easily be set up as profit centers and can then perform their duties just like an outside, market-competitive service provider.

Should it not be possible to operate some profit centers profitably, then, especially in case of units with a negligible strategic significance for the company, outsourcing them should be considered (chapter 6.2).

5.2.2 Transfer Prices

Calculation of Transfer Prices

Often the design of interaction and business relationships between the profit centers constitutes a big challenge. For each product or service a fair transfer price has to be calculated. Management must create a coherent system that regulates the mutual compensations between the individual profit centers. This system should be aligned with the global corporate goals with regards to strategy and value creation. Transfer prices can be calculated in five different ways:

1. **Cost Plus**

    The transfer price contains all fixed and variable costs in addition to a flat profit margin.

2. **Full Costs**

    Full costs comprise all fixed and variable costs; a profit margin is not included.

3. **Variable Costs**

    The fixed and overhead costs are neglected. The price solely compensates for the variable costs incurred by producing/rendering the product/service.

4. **Marginal Costs**

    The price of the product or service transferred covers the additional costs that are incurred for the production/rendering of this additional unit of a product/service.

5. **Market Prices**

The market price of a product or service is determined by supply and demand. The costs are not taken into account.

Independently of the calculation method, the transfer prices can be set in different ways:

- **Bilateral Agreements Between the Units Involved**

  The supplying and receiving unit institute agreements about the terms and conditions, including scope and price. These contracts should contain market parameters, such as the quality of the service, timeframe, as well as sanctioning rules in case of non-performance, and a conflict resolution framework (e.g., through a complaint or arbitration authority).

- **Multilateral Agreements Between Several Units**

  Same as above. However, here the agreements are not just between two but between several units of service providers and receivers.

- **Set-up of a Central Department**

  A staff position created for this task or a special authority determines the current transfer prices at regular intervals. The prices could alternatively be set by an external middleman (for example by a certified public accountant).

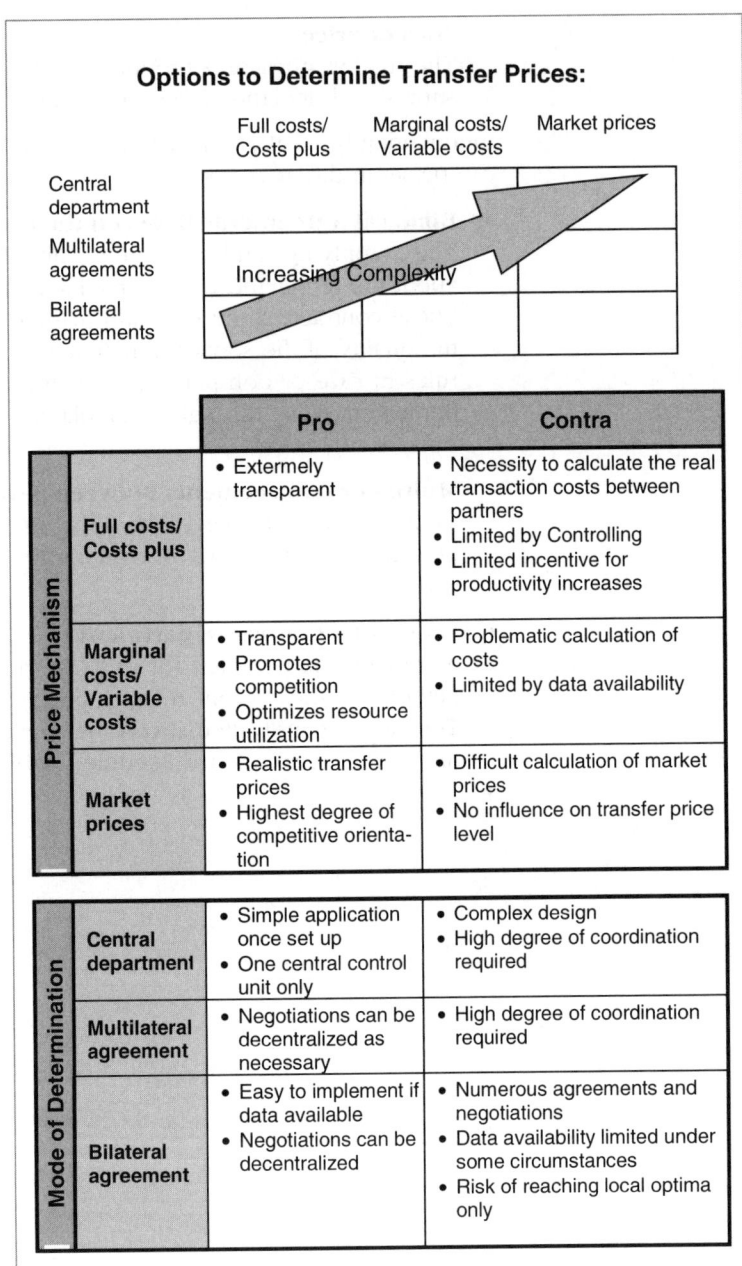

Figure 5-1: Transfer price matrix

Figure 5-1 illustrates possible permutations between calculation methods and set-up modes of transfer prices with their respective advantages and disadvantages. While theoretically the calculation according to market prices is economically the most reasonable method, it is also the most complex one; the optimal system in practice depends significantly on the individual situation. Often, comparable market prices are not available, especially in case of products and services for which a clear-cut external market does not exist.

In addition, the availability of data necessary for the calculation of costs incurred by the business units differs between companies. Also the number of units that interact among each other varies. There is a wealth of possibilities for the development of a value creation - oriented transfer pricing system. Therefore there is no universally applicable procedure – every company has to find a system to suit its needs.

5.3   Metrics as Performance Indicators and Benchmarking Standards

When evaluating the performance of individual units, not only the organizational regulations of the relationships among these units take on a significant role. Also, the employees within these units must be managed through the appropriate metrics.

5.3.1   Metrics for Value-Based Management

Quantitative Indicators as the Basis of a Goal System

The core of a goal system directed towards value creation consists of quantitatively measurable indicators that reflect the financial and strategic position of a company. These indicators consequently serve as performance parameters for the individual departments and mangers in charge. In figure 5-2 the advantages and disadvantages of possible metrics are analyzed.

A control function driven solely by profit is always insufficient, because it neglects growth. A control framework based on value creation is better, but also more complex. It becomes more difficult, the more realistic the value created is represented within the metrics and incentive systems.

Metric/ control function	Advantages	Problems and Disadvantages
Profit (e.g. ROS, EBIT)	• Relatively easy to calculate, evaluate and communicate	• Does not account for capital costs • No control of/ reward for growth and innovation
EVA	• Takes capital intensity into account	• Factual capital costs often hard to calculate • Investments into future markets are not sufficiently considered • Innovational capability is insufficiently accounted for • "Milking strategy" can create EVA short-term, but impairs long-term corporate success
Economic Profit Plus	• In contrast to EVA: takes into account long-term effects through savings rates or inclusion of an innovation rate	• Differentiation between levels and departments required under some circumstances • Complex and too little transparency
Shareholder Value	• Control metric that is closest possible to created company value	• Difficult to implement in practice, especially in case of private companies • The success of a public company is not always due to the unit of the executive position that is to be evaluated/ rewarded • No reasonable regulative against market-dependent over or under valuation – "roulette"? • "Milking strategy" cannot be avoided if unrealistic goals drive up the share price in the short term and shares held are sold immediately

Figure 5-2:   Measuring and Controlling Methods

Controlling by EVA aims at increasing profitability. For companies in dynamic industries or for business units active in strongly growing markets, steering by this metric can be fatal. It includes the cost of investments and acquisitions in the target metrics, thereby degrading them. EVA therefore hardly offers an incentive for investments in new technologies and promising markets. One of the results is that the growth rate in such companies lags far behind the true potential. If these companies fail to design a management system that controls directly by value creation contribution, they need to revert to a method like Economic Profit Plus (that principally builds on EVA). However, in this case the EVA parameter is supplemented by additional components that measure innovation capability (e.g., innovation rate).

5.3.2 Individual Performance Indicators for all Employees

Evaluate Employees per Individual Value Contribution

There is a multitude of possible indicators to measure individual performance. In principle all employees should be evaluated as closely as possible according to their concrete value contribution, so that the individual ambitions can be aligned with the company's objectives through the appropriate incentive systems.

The aforementioned financial indicators are quite a significant factor in the evaluation of executives with budgetary responsibility. But room for qualitative performance indicators should also remain on the highest rung of the ladder. If evaluations are purely based on quantitative factors, managers have little incentive to take the time to talk with staff, become involved in recruitment sessions or take care of public relations. These qualitative measures are effective in the long-term and therefore cannot be reflected in short-term financial indicators. For this reason they have to be explicitly included in the performance measurement.

But even below the level of a department head, employees have to be evaluated according to targets whose realization they can influence. This category includes productivity indicators like machine utilization and value added per employee or quality indicators, such as error rates.

Approximately six to eight performance indicators should be determined per employee. There are examples of companies that take into account 20 or more elements in the performance evaluation process. In such cases it is rarely possible for the individual to focus on the really decisive factors.

Figure 5-3 identifies typical quantitative and qualitative performance indicators for management levels:

Intermediate Target	Metric
Quantitative (financial)	• EVA/ROCE • ROS • EBIT • Cash Flow
Quantitative (non-financial)	• Revenue growth rate • Market share per product • Innovation rate • Fluctuation • Value added per employee
Qualitative	• Quality/customer satisfaction • Employee satisfaction

Figure 5-3:   Typical metrics

In practice there are a large number of activities worth supporting that can only be captured indirectly but which are decisive for sustainable value creation. For this reason a system of metrics and incentives should also leave room for components that only indirectly impact value creation.

5.4   Incentives to Reach Target Metrics

Incentive Systems as a Motivation to Reach Goals

The management can only implement company goals if it succeeds in motivating the members on all levels of the organization. Individual ambitions thus have to be aligned with the goals of the company. Ideally, a win-win-situation is created in which the employees strive for the realization of company goals by pursuing their own interest.

With the appropriate design of the incentive systems, individual ambitions are managed and channeled so that they maximally support the selected value creation levers. In order to serve as performance indicators the target values have to be designed in such a way that the employees can directly influence their realization. If higher performance is not translated into reaching higher goals and thus higher rewards, the entire incentive effect is lost. The head of sales, for example, can be evaluated according to metrics like sales or customer satisfaction. However, he/she does not have any influence on reducing procurement costs. In general the reward system should always maintain and promote performance and competitiveness.

Incentive systems have to be fitted to the respective target value. If, for instance, manufacturing pursues the goal of reducing scrap by 25 percent, then the rewards should reflect that objective. An incentive system that pays the workers solely according to output will tend to oppose this target value. In contrast, a bonus for every percentage point of reduced scrap will stimulate workers to even exceed the target.

In addition, the incentive system also needs to include parameters like quality and employee and customer satisfaction, and structure rewards accordingly. The balanced scorecard method is available as a management instrument that accounts for both the quantitative as well as qualitative aspects – chapter 6.4 will present it in more detail.

Monetary incentives like salary, bonuses or stock options play a central role in almost all incentive systems. In addition, non-monetary components also supplement these incentive systems. Among these elements are, for example, promotion options, company cars, availability of company products at reduced prices, public honors and other privileges. The associated increase in internal and external prestige and influence often ranks as a significant motivating factor. Often even the reputation of a company itself is an incentive for the employee.

During the Internet hype and the related boom in the high-technology sector, stock options were regarded as universal currency to be issued to employees instead of a regular salary. As long as stock prices soared, the recipients were more than satisfied. But this willingness to accept options decreases in times of high capital market volatility, especially when many stocks are excessively losing value. Such option programs can thus lead to added employee fluctuation.

Companies can even reduce their market value by issuing stock options too generously. If employees with a relatively low base salary receive a great number of stock options over a longer period of time, they redeem their options or sell stocks held in case of liquidity problems. The departure of co-founders who own significant share packages or other upheavals within the company can also perpetuate the sale of stocks. A sensitive stock valuation can thus come under even more pressure.

It follows that the design of the incentive system will often determine the success of a value creation program, since employee performance is driven by their motivation. Because of the great significance of this aspect, it is discussed in detail on the level of the individual in chapter 6.4 (human capital).

6

The Organizational Foundation:
The Basis of the House of Value Creation

This chapter describes the foundation of the House of Value Creation. It presents real value creation potentials in the areas of portfolio structure, business architecture, corporate finance, human capital, and investor communications.

6.1 Portfolio Structure

Strategic Planning and Portfolio Management

Portfolio & group structure and the related degree of flexibility, freedom, and transparency for capital markets impact the market valuation of a company. Especially conglomerate structures, which are too tightly interwoven and afflicted with potential for cross-subsidization and other measures that restrict competitiveness, lead to substantial discounts on the capital markets. Investors expect companies to either restructure or liquidate loss-bearing business units instead of allowing them to hurt earnings in the long term. Excessive vertical integration and great differences between the business architectures of individual business units can also have a negative impact on market valuation. Few conglomerates – among them General Electric – attain group valuations that exceed the sum of the valuations of their individual business units. Consequently, enterprises should align their portfolios and organizational structures towards maximum flexibility and transparency. In this context, strategic planning and portfolio management play crucial roles (box 7).

Box 7: Strategic Planning & Portfolio Management

Strategic planning is the process of adapting corporate objectives and resources to the changing market environment. Different industries and business segments produce different margins. Consequently, a company must manage the portfolio of strategic business units so that it creates and maintains success potentials, which result in increasing profits.

Strategic business units are autonomously managed corporate sections (i.e., product-market combinations) that are independently positioned on the market to secure the long-term survival of the enterprise as a whole. A separate strategy must be developed for each of these units. All product-market combinations must be clearly separated from each other and target the development of competitive advantages.

Portfolio management is a crucial part of strategic planning and aims at creating sustaining longer-term competitive advantages. Portfolio management determines whether strategic business units are to be maintained, developed or relinquished. The development of new strategic business units can be achieved internally through organic growth or driven by corporate mergers.

Portfolio management begins with a portfolio analysis, which maps and evaluates a company's own products within the competitive environment. Complex interrelations are simplified and quantified by consolidating the factors that influence them. Often, strategic business units are plotted in a matrix, which separates the factors that are within and beyond the company's control. This presentation aids in recognizing company-specific strengths and weaknesses as well as market-related opportunities and threats to the individual strategic business units early on.

The two most widely used portfolio analysis methods are the market share-market growth portfolio and the market attractiveness-competitive advantage portfolio.

6.1.1 Strategic Portfolio Optimization – Increasing Transparency

Effective portfolio management requires permanent monitoring of all business units. Divisions that may still be considered part of the core business but which are heterogeneous in terms of industry clockspeed, business architecture, employees, or customers should be managed as independent units. The capital market may very well reward the spin-off of individual business units with a premium, provided transparent organizational structures are in place. After spinning off Lucent and NCR in 1996, the U.S. telecommunications giant AT&T was able to increase its market capitalization from US$75 billion (AT&T) to US$159 billion (AT&T, Lucent and NCR) in just two years. In fact, on the day the spin-offs were announced, the share price rose by more than 10 percent.

There are numerous reasons for such *conglomerate discounts* in valuation. Conglomerate business units report to their executive management, not to the capital markets. Moreover, they are usually managed according to common group metrics rather than based on the individual industry clockspeed and their growth horizons (chapter 4 and 5). This fosters inefficiencies and infighting for resources and consequently creates local optima rather than a holistic alignment towards value creation. The lack of transparency at conglomerates is also responsible for the continued existence of unprofitable business units at the expense of profitable ones - to the detriment of the overall capital market valuation.

Companies generally have three options to increase the transparency of individual business units in the eyes of the capital markets:

Spin-Off

In the course of a spin-off, a business unit is usually placed on the capital markets as a legally and economically independent entity. While this improves the company valuation, it also provides the parent corporation with additional resources for the funding of growth options. There is no need for the newly established enterprise to remain connected to the former parent in any way. The parent company's existing shareholders divide the stock options for stakes in the new company among themselves. The new, spun-off business, however, is often no longer compatible with the stock portfolio of the existing shareholders. Instead, it is much more attractive for a new group of investors, who would then also be prepared to value the spin-off higher. For instance, in July 2000, Cabot Corporation, a global player in the chemical industry spun-off its entire semiconductor segment Cabot Microelectronics (CCMP). During the allocation of shares held by the parent company to existing shareholders after the initial public offering (IPO) in April 2000, the price of the stock had already gained 140 percent.

Partial Initial Public Offering

The term partial initial public offering (partial IPO) refers to the partial sale of equity capital on the capital market. Because only a portion of the subsidiary's equity is sold, this does not, however result in complete independence. Nevertheless, it does increase the autonomy of a business division, since its management gains access to its own fixed assets. As a rule, the parent company usually retains a majority stake in the new subsidiary, allowing it to continue to exert influence and to participate substantially in the success (or lack thereof) of that subsidiary. DuPont spun off 30 percent of its oil business, Conoco, in 1998, which amounted to the largest equity capital sales event ever at that time. The objectives were to foster Conoco's future growth through independent access to the capital market and to increase management flexibility. Since early 1999 the company was consequently able to increase the company value of America's fourth largest oil corporation by almost 50 percent.

Tracking Stocks

The term tracking stocks refers to the issuance of a new security for an individual business unit by the parent company, which is then valuated by the capital market. This security reflects only the success of the division, not the performance of the overall group. Contrary to the first two options, this method does not call for the legal and institutional autonomy of the business unit. Consequently, there is no need for an expensive endowment with its own capital asset base. On the other hand, the differentiation of the business unit from the parent company is only marginal. Through the tracking stock method, US telecommunications provider Sprint was able to triple the capital market value of its mobile phone sector (Sprint PCS Group) between the first day of the offering (end of 1998), and September 2001.

Spin-offs, partial initial public offerings and tracking stocks allow shareholders who are not interested in the parent company to invest in selected business units. Practical experience has shown that capital market differentiation may very well attract new investors. When the American telecommunications provider US West spun off its Media Group via tracking stocks in 1995, 86 percent of the new business units' shareholders were new investors.

6.1.2 The Generation Conflict – The Daughter Becomes Independent

Costs and Challenges of Portfolio Optimization

The independence of a business division entails organizational challenges. Successfully mastering these challenges is a prerequisite for the creation of real added value for new and existing shareholders. Consequently, corporations are well advised to also apply the House of Value Creation principles and to implement comprehensive value creation programs to newly autonomous business units and individual parts of the conglomerate.

The units separated from the parent company must be clearly defined in terms of their autonomy and their relationship with the parent corporation. For a spin-off of business units to be profitable for both parties involved, it is usually essential that the new unit does not compete with the parent and that the parent's remaining divisions do not cannibalize the activities of the newly created subsidiary.

In most cases, parent company and subsidiary remain closely intertwined even after the institutional separation has been implemented. These interdependencies must, however, be transparent and clearly regulated. In the event that one of the two companies purchases goods or services from the other, a fair transfer pricing system must be set up. This agreement should not pertain exclusively to pricing, but also govern the long-term relationship between parent and subsidiary. It would be fatal if the true success/failure of parent and subsidiary were obscured by nontransparent transfer pricing. It is of the utmost importance not to put the capital market's trust in the plausibility of the independent subsidiary's figures at risk through arbitrary profit (and loss) transfers.

The management of the interdependencies must take the complicated relationship between group parent, independent unit, and capital markets into account. If the parent retains a stake in the independent unit, this triangular relationship becomes even more complex. Every time the parent exerts its influence – in particular when it changes its equity stake in the independent unit by repurchasing or issuing new shares – it encounters a conflict of interest between its own shareholders and those of the subsidiary. Aside from potential legal issues, there is always the danger of losing shareholder trust and consequently destroying substantial value.

In conglomerates, individual business units are frequently treated as stepchildren by their own management and are therefore furnished with insufficient resources. This prevents them from being successful in competing with independent competitors. On the other hand, some units might claim too many resources of the group, thus withdrawing capital that could be more effectively used in other areas. Separations of business units usually always solve the problem of sub-optimal resource allocation, since now resources are no longer allocated during budget negotiations but have to be earned on the capital market.

The separation of business units on the capital market can strengthen the parent company's position during corporate takeovers, thus enable external growth. If the parent's shares are valued below that of potential takeover candidates in terms of price/revenue or price/earnings ratio, they are hardly a suitable currency with which to go shopping. In such cases the anticipated higher-valued shares of independent business units might represent a better means of payment.

Portfolio optimizations have the potential to create substantial value, but there are challenges that must be reckoned with. The more of the key questions below can be answered with a definite yes, the better the chance of a successful separation:

- Are the business activities of the unit outside of the parent's core business – or even in another industry?

- Is the unit simply integrated into the group? Is this type of integration hardly justifiable?

- Is the individual success of the unit non-transparent and is the unit insufficiently visible on the capital market?

- Does the unit have problems in recruiting executives or qualified personnel?

- Is the business model of the unit not compatible with the conglomerate's model (customers, industry clockspeed, and capital intensity)?

- Does the unit grow more quickly or slower than the conglomerate?

- Do competitors of the unit who are not part of conglomerates attain different valuations than the unit?

6.2　Business Architecture

The business architecture encompasses the corporate organization, processes and the IT infrastructure. It defines which employee performs which function, for what purpose, at what time and location with what methods, to reach the company's objectives. In other words, the business architecture describes how the company creates value.

6.2.1　Optimization Levels – Organization, Processes and Systems

The organization can be divided into a structural and an operating component, and it is sub-divided into processes. It defines the business infrastructure and stipulates the principles based on which the entrepreneurial activities are to be controlled. It encompasses the creation and coordination of functional areas as well as the implementation of the required procedures. Processes detail the technological, temporal, and local interdependencies of factors, which lead to the realization of specific goals at a specific level of quality. Last, but not least, the systems define all technical facilities required for information acquisition and processing.

Design of the Business Architecture

There is a natural correlation between organization, processes and systems (i.e., the IT infrastructure): The organization defines the required processes. These in turn determine what systems are necessary to map them. Consequently, an optimization of the organization always requires the adaptation of the underlying processes, too. Modifications to the processes, on the other hand, also impact the systems. Measures aimed at improving efficiency therefore should not approach these three levels from an isolated perspective (figure 6-1).

Due to internal structural changes and the progress of globalization, more and more companies are being confronted with becoming larger and more complex. This brings with it a need for constant optimization in order to be able to react flexibly to changing market conditions.

Given the increasing speed of information exchange, extremely dynamic and global markets are also immensely competitive. Competitive advantages can often only be attained or sustained through continuous optimization of the business architecture. Nonetheless, the competition catches up quickly. Consequently, profit margins erode and are decreased even further by the costs of continuous investments.

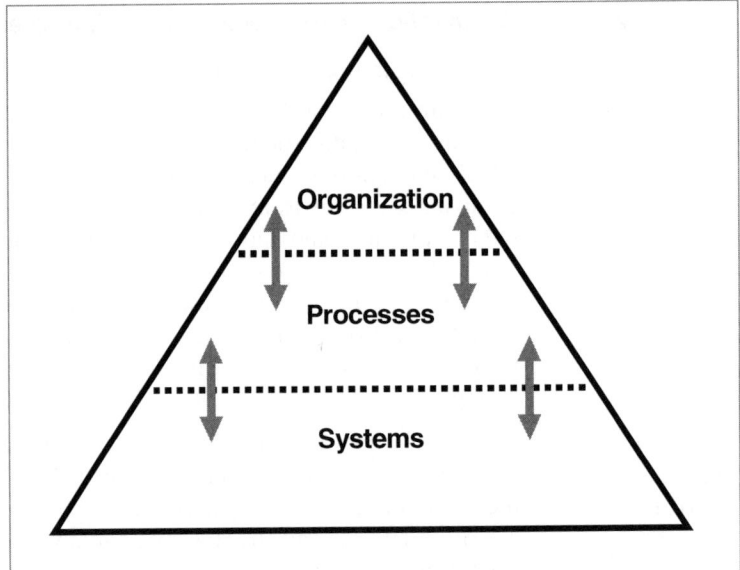

Figure 6-1:   Correlation of organization, processes and systems within the business architecture

6.2.2   Optimization Approaches

Prior to optimizing the business architecture, a detailed analysis of the current business situation has to be conducted. On this basis, improvement strategies can be developed and concrete, targeted optimization measures selected. Chapters 3 and 4 describe relevant approaches for such company analyses in more detail.

The starting point for the selection of optimization options should consist of functional (or process) cost pools that are clearly separable along the organizational or cost dimension (e.g., cost centers). Responsible managers have to be assigned for each optimization to be initiated. A value creation program identifies the cost pools to be optimized and determines the targeted approaches with which to realize these optimizations. Figure 6-2 shows the significant cost pools and assigns examples of optimization measures.

Below, various typical optimization measures for the different cost pools are explained. The list is by no means exhaustive but represents a selection of proven and widely used optimization measures. The goal is to elaborate on the value creation program in more detail. Additional and more specific descriptions of these measures can be found in the relevant specialized literature.

Each measure listed has a defined objective and a typical area of application. Nonetheless, some of the measures are transferable to other areas, with or without adaptations. Outsourcing a section of the company's value chain is a measure that can be principally applied to all corporate functions, not just to production. The responsibilities of the human resource department, for example, can be outsourced to a personnel management service provider, such as e-Peopleserve. Due to the interdependencies described earlier, no measure should be considered an isolated act that affects only one level.

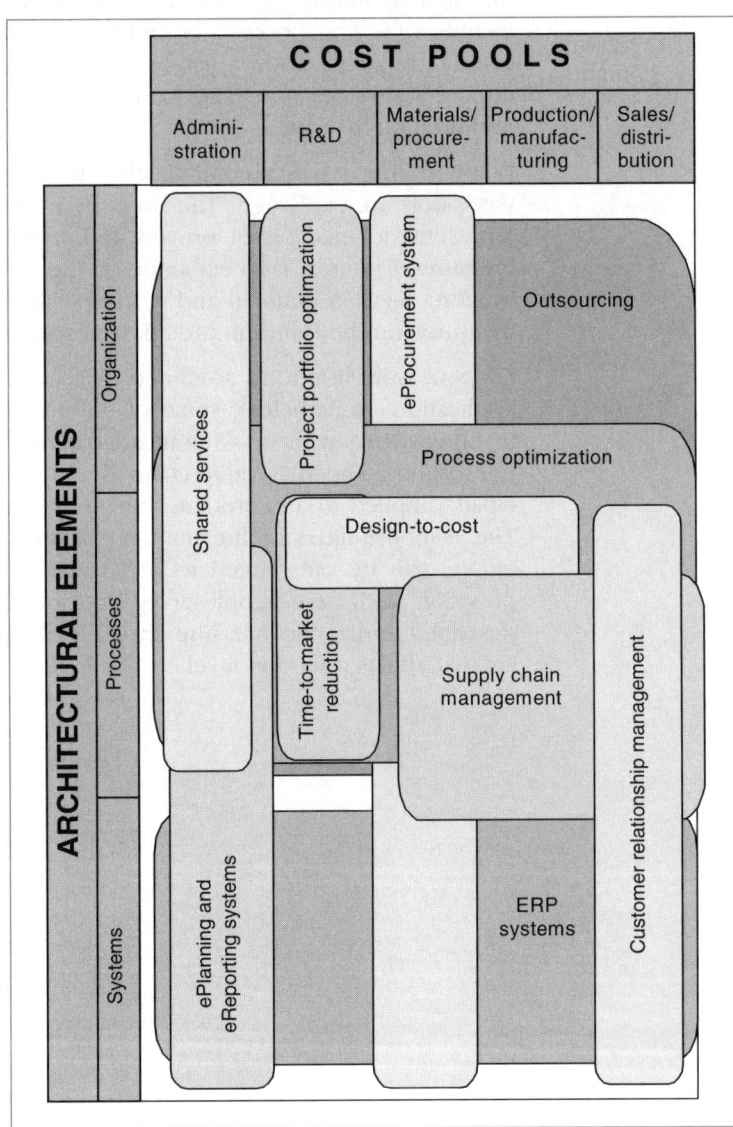

Figure 6-2: Selection of optimization measures

6.2.2.1 Process Optimization

Process optimization aims at the identification and realization of savings potentials in operational processes. The objective is to reduce costs in the areas evaluated through improved effectiveness and efficiency. Non-value-adding activities are terminated, interfaces optimized, and processes simplified. At the same time the quality level is improved and throughput times are being reduced. The simultaneous optimization of the three dimensions cost, quality, and time may appear contradictory at first, but it is not. The optimization of preventative maintenance procedures (Total Productive Maintenance – TPM), for example, can also yield lower repair costs, avoid scrap, and reduce idle times.

Process Optimization to Improve the Competitive Situation

Process optimization denotes a methodical guideline for the formulation of streamlining strategies aimed at improving a company's competitive positioning. Using creative techniques, cost reduction approaches are determined based on a cost/benefit analysis for each individual service. The ideas that are developed during this process are subsequently prioritized according to impact and feasibility, evaluated in detail, coordinated with those involved, and then ratified for implementation.

In addition to the cost dimension, process optimization also takes quality and time into account. Generally this method is applied to the areas production/manufacturing, and administration, but it is not limited to these departments. An efficiency analysis is contingent upon the technical and organizational boundaries. To this effect, everyone involved must commit to the ambitious goals. The process, which has been tried and tested in practice, is shown in figure 6-3:

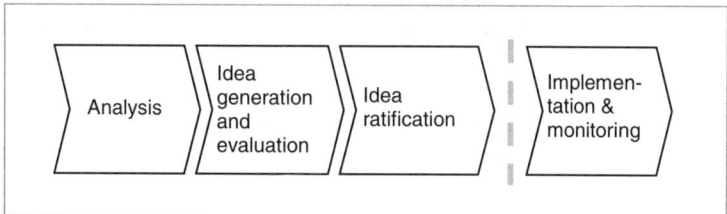

Figure 6-3: Process optimization phases

Phase 1

Analytical Phase
For the company areas to be analyzed, all costs incurred and services rendered are recorded. In order to determine the improvement potential this analysis must differentiate between costs that are within and those that are beyond the division's control. From those costs that can be influenced, an improvement potential can be derived. Experience has shown that this potential usually amounts to a double-digit percentage value.

Phase 2

Idea Generation and Evaluation Phase
In the idea generation and evaluation phase employees generate as many ideas as possible in a bottom-up procedure. The predetermined minimum improvement target should be high (in practice it is often around 40 percent) in order to stimulate creativity for the idea generation process and to obtain truly meaningful results. Upon completion of the idea generation process, the ideas are evaluated by those affected. Three idea groups are then categorized according to realization and economic benefit criteria: Go, Check, and Drop ideas (figure 6-4):

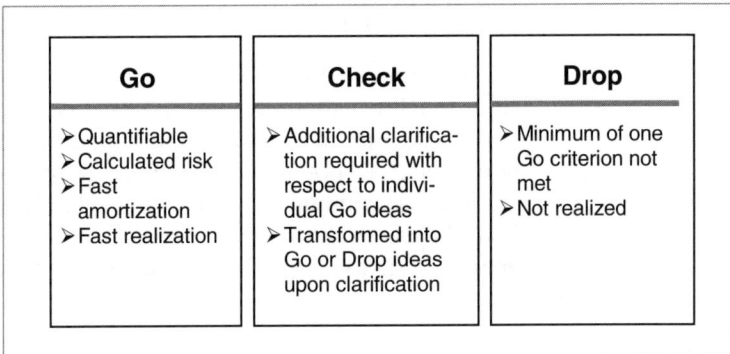

Figure 6-4:   Idea evaluation and selection

Go ideas are being considered for realization. They are quantifiable, i.e., their results are measurable. A company implementing them only runs a manageable risk. Implementation costs are typically recovered rather quickly. *Check* ideas require further clarification regarding one or more classification criteria. Depending on this clarification, Check ideas are re-classified as either Go or Drop ideas. An idea that does not meet one of the

Go criteria is classified as a *Drop* idea and is consequently not considered for realization.

Idea Ratification Phase

Once evaluated ideas are ready for ratification, a responsible steering committee selects those ideas that are to be realized. A realization plan is only compiled for category Go ideas. This plan defines who is responsible for the implementation of the identified improvement measures, in what time frame the implementation is to be carried out, and whether investments could be required. It also estimates the impact of the measures on the company's performance.

In 1994/95, for instance, IBM's semiconductor division executed a process optimization program aimed at reducing costs at its in-house production plants. The comprehensive worldwide optimization of the production processes lowered total production costs by up to 25 percent. Variable production costs could even be reduced by up to 50 percent. Countless other companies have been successful in bringing down the administrative costs by systematically applying process optimization programs.

6.2.2.2 Creation of Shared Services

The creation of centralized services, which are also referred to as shared services, is a method that has proven effective in reducing administrative overhead costs. The shared services concept consists of establishing a competitive, service and market oriented organizational unit with full profit and loss responsibility. This unit takes over processes that are transaction-based and can be standardized or perform support functions, and offers them to internal customers. This allows operating business units to focus on their own core competencies. Shared services are principally suitable solutions for all repetitive, quantifiable internal services. Due to consolidation and standardization, shared services realize economies of scale and yield cost savings through improved utilization of resources and systems.

Shared Services as Starting Point for Process Optimization

The consolidation of services into shared services is frequently the starting point for a comprehensive process optimization and standardization program. Typical projects attained cost savings between 20 and 40 percent in conjunction with the launch of shared services. As a rule, the consolidation of individual functional areas and the resulting reduction of interfaces and redundancies accounts for some 50 percent of these savings. Standardization and process re-engineering are usually responsible

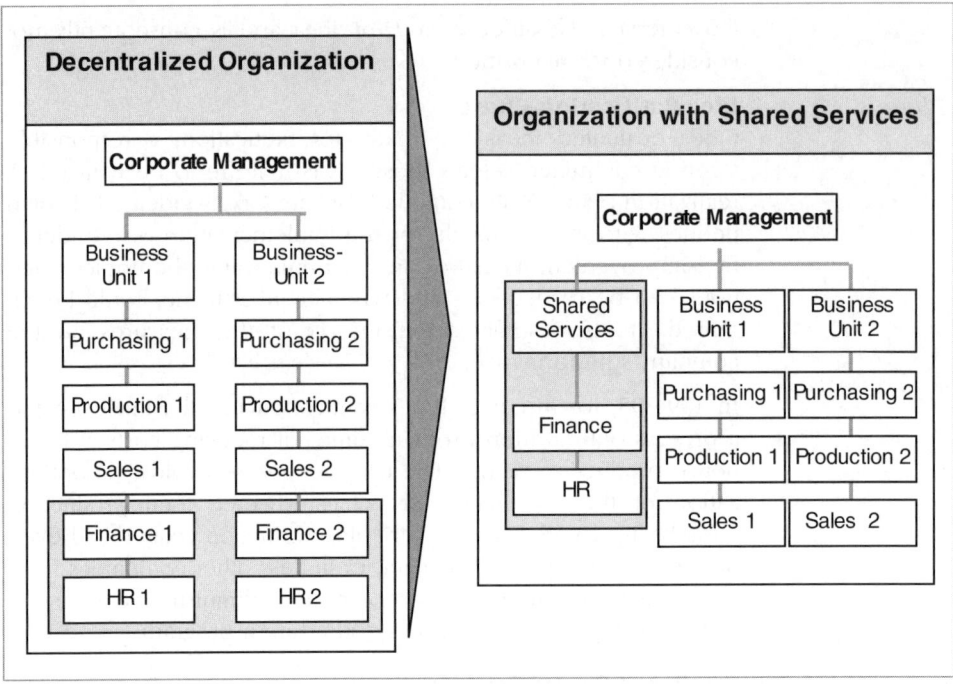

Figure 6-5:   Creating shared services

for about 25 percent each. In other words, shared services offer a compromise between centralized and decentralized organizational structures (figure 6-5).

Obviously, this type of organizational structure does have some disadvantages. Given that the structure is much more complex in terms of communication and transactions between the shared services centers and the other corporate units substantial costs are often incurred. The harmonization of the existing IT-infrastructure required to accommodate shared services can actually be cost-prohibitive. Sometimes the creation of a shared services business unit also fails due to the resistance of affected employees.

Because their processes for the most part are easily standardized, the finance, IT, purchasing and human resource functions are particularly well suited for the creation of shared services. Generally, however all corporate functions that meet the criteria of marketability and company-wide demand are suitable for the establishment of shared services. A market-oriented price based on supply and demand interactions must be able to form and scale effects must exist. Only then can consolidation of functions into an independent organizational unit yield substantial cost benefits.

Shared services make internal services transparent and are driven by the (internal) customer's needs. Costs are transformed into prices. Thanks to the achieved transparency, the quality of the services rendered can now also be controlled and adjusted if required. The transparency of the service portfolio allows comparisons with external vendors. Another possible scenario would be the spin-off of shared services. In this case the services portfolio could also be offered to other companies.

A consumer goods company consolidated its accounting activities at 15 different locations on a Pan-European level. Five shared service centers with cross-country responsibilities were formed, yielding cost savings in the amount of US$24 million, or 21 percent of this function's total costs.

One of the world's largest gas producers with production facilities in 60 countries was facing multiple challenges in 1997. The company grew only moderately. Costs were too high in comparison to the industry average, especially in non-production areas. The company valuation fell below the value of its fixed assets, turning the company into an ideal take-over candidate. The executive management decided to implement a radical cost reduction program. In addition to launching a production process optimization program, shared financial services were established as well. The standardization, consolidation and automation of functions distributed throughout more than 20 locations into three shared service centers yielded cost reductions of 35 percent in the area of accounting. The successful implementation of the overall program increased the market value of the company by an impressive 200 percent.

6.2.2.3 Introduction of ePlanning / eReporting

For corporate management, first-class planning and reporting instruments are of crucial importance as they provide the executives with real-time data for all finance-related decisions. In addition to this central executive management function, the reporting systems must also facilitate the timely compliance with statutory requirements, such as the compilation of annual reports.

An efficient ePlanning and eReporting system (later referred to as eReporting system) significantly reduces the time and costs that have to be expended for the provision of necessary financial data. The procedure of introducing such a system consists of three phases: Analysis, design and implementation (figure 6-6).

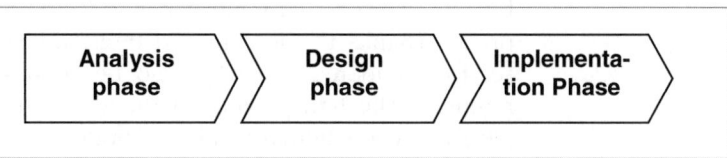

Figure 6-6:   Phases of an eReporting system implementation

Phase 1

Analysis Phase

In a first step, the company's planning and reporting process is evaluated. The new planning and forecasting process is then designed based on the derived requirements.

Phase 2

Design Phase

The actual system is created in the design phase. Here, functional requirements need to be taken into account with respect to reporting, special demands of the executive team as well as general system requirements. The goal is to reflect all requirements in a single consistent system architecture.

Phase 3

Implementation Phase

Once the required system has been designed, it is implemented in the last step. During this phase, a standardized eReporting system is tailored to the specific company needs, and tested for compliance. Modern eReporting systems are usually integrated into the company-wide *ERP system* and are web-based. This makes the introduction of isolated applications redundant. For a full discussion of such a system, please refer to Efficient eReporting with SAP EC - Strategic Direction and Implementation Guidelines, by Andreas Schuler and Andreas Pfeifer.

The procedure of installing an eReporting system and the associated cost reduction potential are demonstrated along the example of telecommunications provider Tenovis (box 8).

Box 8: Tenovis Case Study

Tenovis was formerly the "Private Networks" business segment of Bosch Telecom. It was acquired by KKR, a private equity company. The number 2 on the German market, Tenovis generates annual revenues of approximately US$1 billion with about 7,000 employees.

After separation from the Bosch Group, Tenovis' infrastructure was too weak to meet the capital markets' and new owner's information requirements. The restructuring initiated by the new management and the establishment of decentralized profit centers called for a wide distribution of entrepreneurial responsibilities. Consequently, several hundred planners had to be involved – all with binding accountability for their particular areas. To this effect the company was looking for a planning system that was capable of decentralized recording of revenue and cost center planning data. In addition, the results of the entrepreneurial operating units had to be consolidated into profit and loss accounts on a per-business-unit basis. The only solution that made sense was a web-based planning system. This way, a large number of users could be reached with minimum rollout investments.

A substantially simplified, parallel planning process was designed based on existing planning processes and crucial planning content for the businesses analyzed. Aligned to this process, a new and also decentralized forecasting process was created and the planning/results calculation was analyzed and defined. This allowed the mapping of the changed processes onto the new company structure with several functionally independent business units that carry their own profit and loss responsibility. The cost center landscape was also re-defined and a simplified, automated procedure for the distribution of overhead expenses was created. The functional requirements of the process and new administrative methodology were translated into a system design. This in turn formed the basis for building the decentralized (web-based) planning component and the centrally used analysis component. The latter component of the system was defined as the data basis for a reporting method that provides planning data based on the current corporate structure and is capable of acquiring the pertinent as-is information.

125

At Tenovis, the introduction of an eReporting system delivered substantial improvements:

- The planning cycle was shortened by 50 percent.

- The restructuring of the method for calculating operating results yielded a decrease in planning complexity. The number of cost centers was reduced by 30 percent thanks to the elimination of clearing account cost centers.

- Thanks to a simplified allocation process, the time and expenses spent on the compilation of the plan/actual calculation was reduced by 25 percent.

- For the first time, all employees with earnings responsibility could be involved in the planning process. Expanding the number of participants from 30 to 500 significantly increased data quality. Another first: The plan/actual calculation was available for all corporate divisions.

- A uniform basis for the reconciliation between sales and production was introduced for the first time. The consolidated database of the integrated planning system simplified plan validation. The company estimated that it was able to reduce the time and cost for the reconciliation process by 80 percent.

- The launch of the system also marked the debut of a partially automated reporting database, which contained a basis of consistent planning figures for all business units. It was thus capable of recording all information of the as-is reporting and of providing standard templates for a selection of financial data based management reports.

6.2.2.4 Project Portfolio Optimization

Optimization of the R&D project portfolio ensures that individual projects meet expectations – especially in terms of business potential. If this is not the case, such projects should be terminated to make the conserved resources available for other tasks. In this context, a selection criteria catalog is compiled, which provides the tools for a review of each individual project. Depending on the necessary depth of content penetration, the optimization process might call for thorough analysis or several rough estimates. It is possible and also desirable to include iteration loops at all junctions of the filtering process. Examples of widely used filter criteria are shown in figure 6-7:

Filter 1	Is there an actual market for the planned product? If a market does not yet exist at this time, does the product have the potential to create a new and lucrative market? If the answer to both questions is no, the desirability of continuing the development of this particular product should be questioned.
Filter 2	Do the benefits of the product justify the investments required to develop it? If the required investments exceed a reasonable amount, given the company's capabilities, the amortization of the product development could take excessively long.
Filter 3	Does the planned product promise sufficient growth potential? This potential should be consistent with the company's growth requirements in the context of its current capital market positioning.
Filter 4	Is the planned product part of the company's core competencies? If this is not the case, it could still be lucrative for the company to develop a new competency. On the other hand, the product might also be suitable for a spin-off (portfolio optimization).

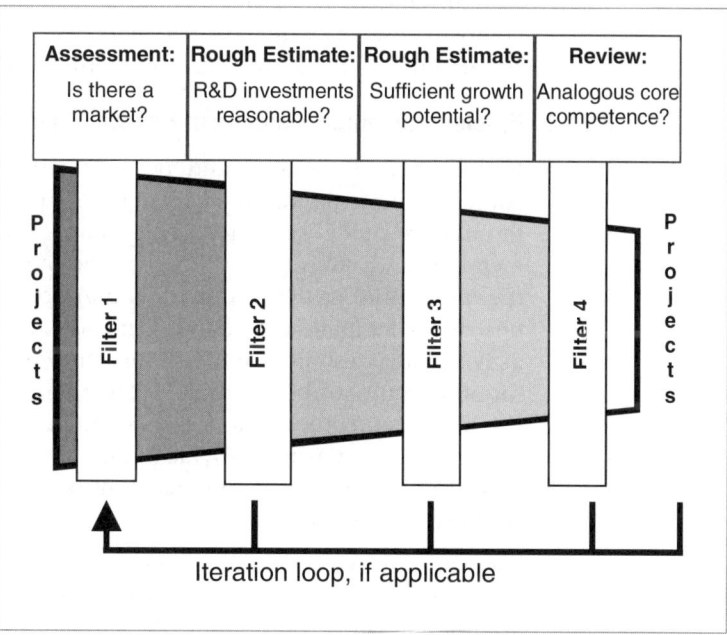

Figure 6-7:   Project selection using a filtering procedure

None of the filters necessarily represents an exclusion criterion. However, if a project does not fulfill one of the filter criteria, this fact provides very helpful alerts to potential problems, which may lead to the adaptation or termination of the project and consequently the allocation of resources to other projects. The reduction of the overall number of development projects through the application of the filter method ensures that only promising projects are realized and that costs are reduced in areas that do not serve the company's objectives. The filter method has also proven its effectiveness in the evaluation of new product ideas, for example within the scope of an internal business plan competition (chapter 4).

The Apple Newton MessagePad constitutes an example of suspending the development of a product that was considered truly innovative but seemed to lack the necessary growth perspective. In 1993, Apple developed the first Personal Digital Assistant (PDA) and thus created a completely new market. Despite a relatively low sales volume, the product was successful during the first few years after its launch. Relative to Apple's overall revenue and to the growth expectations, however, neither the company itself nor the capital market considered the product successful, and it was consequently terminated soon.

6.2.2.5 Efficient Resource Planning in Research and Development

In the area of research and development, the anticipated resource investments required for an R&D project are determined through resource estimates. The more precise the estimation methods and subsequent progress monitoring procedures are, the more efficient the resource planning process will be. This in turn leads to minimized costs. Estimates must take all required activities and resources into account and include realistic factors for all products to be developed. They must also consider deviation factors to compensate for uncertainties and risks. Ultimately, a resource estimate provides the foundation for project-accompanying progress and cost controls - and thus for efficient resource utilization.

Estimates are well-founded prognosis tools that should follow a logical approach and take past experiences into account. Their applicability must be continually verified and may have to be adapted to changing conditions. The actual estimation process consists of four steps and includes a top-down and a bottom-up component (figure 6-8). Estimates are initially developed for individual phases, which are subsequently divided into individual activities and then evaluated. Finally, these activities are divided into smaller tasks, for which estimates are being made. These task estimates are then consolidated into an overall estimate in this bottom-up fashion.

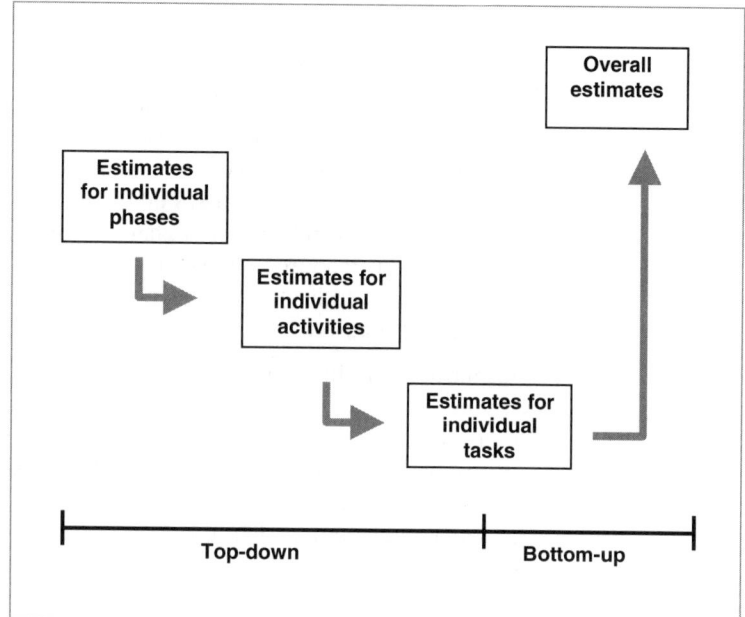

Figure 6-8:   Estimation process methodology

Several competing techniques are currently available for the estimation process, which feature modifications for different industries (e.g., machine construction and software development). The three most widely used estimates are:

- **Estimates based on expert ratings**
  This technique adopts expert ratings as the basis for its estimates. It works well if past project experiences are not available.

- **Comparability based estimates**
  Typically, this technique uses a top-down approach. Estimates are generated by comparing the current project directly with relevant past projects. Such comparisons can

be performed on the project, the component, and the requirement level.

- **Detailed Estimates**
  Typically, this technique utilizes a bottom-up approach and defines each detailed project task. The overall project potential is then derived from the estimates for each task. This technique makes adaptations to changed requirements comparatively easy.

Earned Versus Burned

For the subsequently necessary continuous progress controls, the *Earned versus Burned* method has proven useful. The core idea behind this approach is to compare the costs incurred for a project to the results achieved at that point, whenever predefined milestones are reached. This method guarantees that the remaining expenses are not underestimated. The fact that 50 percent of the planned resources have been consumed (burned) does, however, not indicate that 50 percent of the project has been completed (earned).

Application of the Earned versus Burned method necessitates the setting of sufficiently small milestones that are usually no more than two weeks apart. Only then is this method useful, since incomplete development steps (attributes) are ignored when a milestone is reached (figure 6-9). Based on the status reports, product management has the option to interfere with this process at any time by taking appropriate measures.

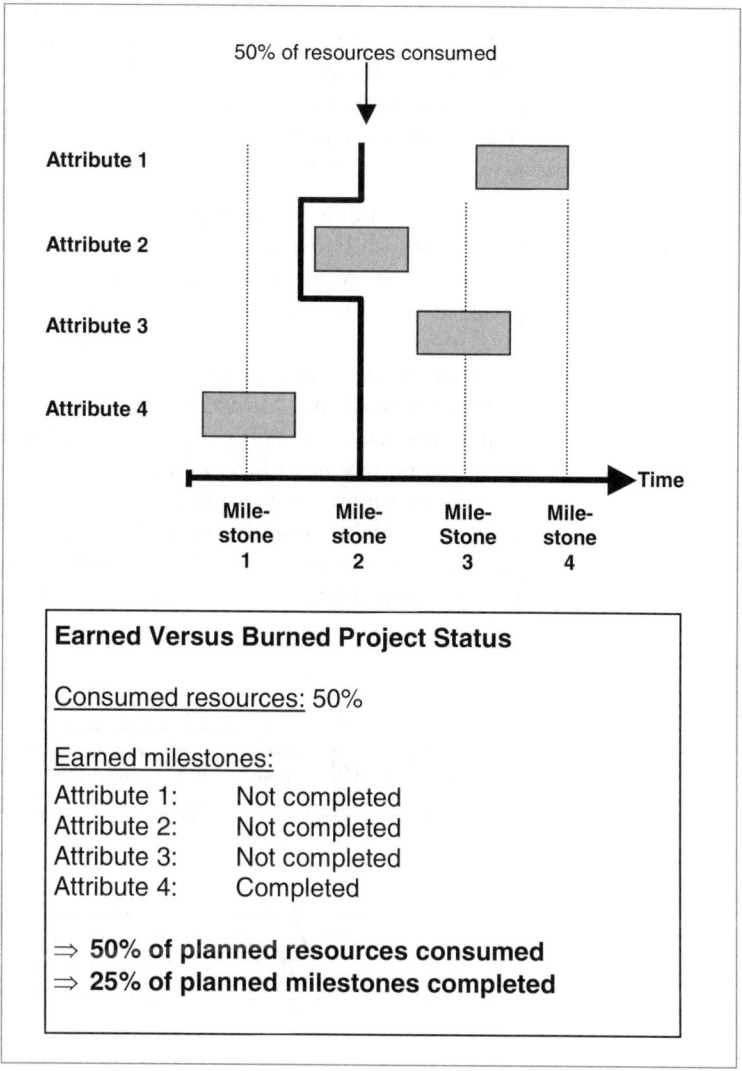

Figure 6-9:   Earned versus Burned method

The example of a medical technology company emphasizes the importance of monitoring achieved objectives through continuous progress control. The company initially monitored the progress of its development activities based on cumulated project costs. It assumed that the partial project results attained were in line with the resources consumed. Consequently, apparent problems in the course of projects were detected too late and the costs exceeded the planned estimate by more than 100 percent. After the company implemented the Earned versus Burned progress control system, it was able to control its progress on a result-oriented basis for the first time. Consumed resources were now compared to the project results actually attained.

Product
Management

An efficient multi-function product management can also optimize costs and resource utilization – not just in research and development. Many companies do not maintain an executive position (product management) that is accountable for progress and expenses, and that coordinates all relevant activities throughout the entire product life cycle. Companies that lack a product management function are at risk insofar as market research results, controlling objectives, and product definition requirements may have little influence on the enterprise's research and development. Instead, the developers tend to frequently aim at what is technologically possible (figure 6-10). Consequently, products often overshoot relevant market requirements, causing unnecessarily high production costs and long development lead times.

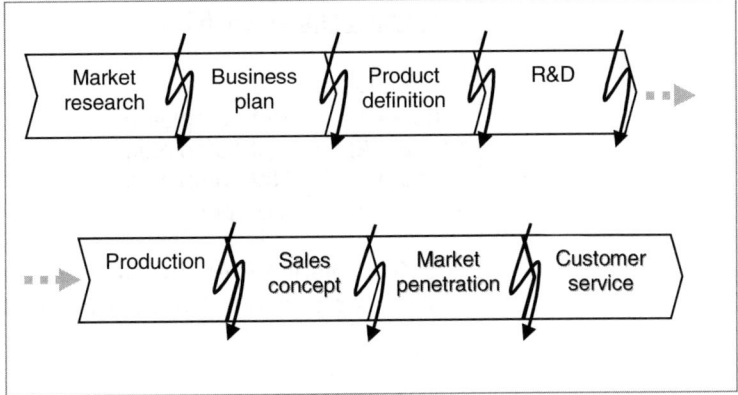

Figure 6-10:   Interface problems due to lack of product management

The implementation of a product management function allows companies to control these risks (figure 6-11). Throughout all phases of the product life cycle, product management remains in charge of the product and the coordination of related activities.

Consequently, product management should be understood as a supervisory measure since it monitors the project's progress as well as relevant quality and cost issues. To this effect, it is not restricted merely to research and development, but takes on the responsibility for the entire product life cycle, from market research to customer service. It increases the efficiency and effectiveness of product development by challenging its activities from an output perspective. In consequence, research and development become based more on product requirements not on the technological feasibility.

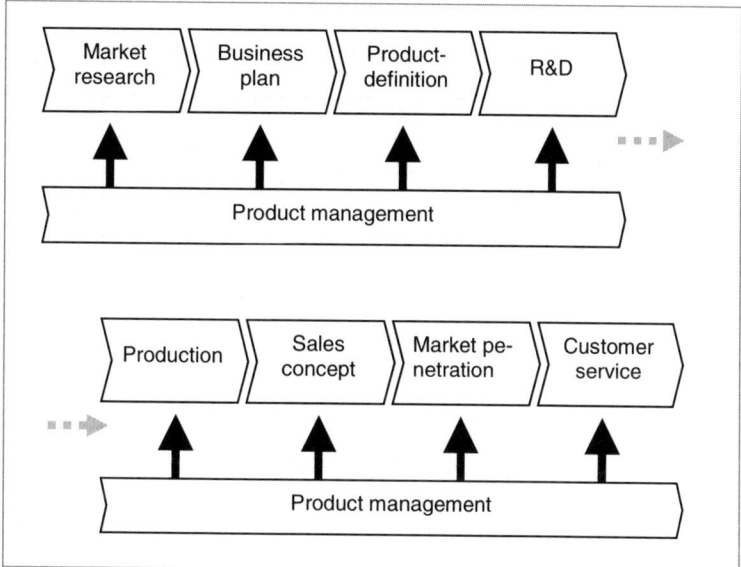

Figure 6-11: Product management involvement and shared responsibility

One consumer products company restricted the responsibilities of its product management solely to those parts of the value chain that followed the sales concept creation but did not integrate the function into preceding parts. This resulted in extreme cost increases, particularly in research and development because no clearly defined goal monitoring was being applied. The R&D department did not focus on the development of the predefined product, but proceeded according to its own priorities. Only after the product management function was allowed to challenge

133

this approach were the development lead times reduced by 50 percent while costs in this area dropped by 25 percent.

6.2.2.6 eProcurement System Implementation

The term eProcurement refers to procurement via electronic market places, industry portals, and auctions on the Internet through networked systems that automate and decentralize processes. Purchased goods and services are responsible for a large share of an enterprise's value creation potential. Consequently, procurement has evolved into a key success factor for businesses and is becoming more and more important. Procurement transactions should be cost-effective, fast, and transparent. EProcure-

Industry	Savings Potentials
Aeronautics	11%
Chemicals	10%
Coal	2%
Communications	5-15%
Data Processing	11-20%
Electronic Components	29-39%
Food-Related Products	3-5%
Wood	15-25%
Transportation Services	15-20%
Healthcare	5%
Pharmaceuticals	12-19%
Machine Components (Metal)	22%
Media & Advertising	10-15%
MRO	10%
Oil & Gas	5-15%
Paper	10%
Steel	11%

Source: Goldman Sachs, 1999

Figure 6-12:   Material cost savings potentials

ment targets precisely these objectives and impacts all cost-drivers in procurement, i.e., purchase prices (material costs) as well as procurement process expenses. According to a 1999 Goldman Sachs study, the industry-dependent procurement process cost savings potential ranges from ten to 25 percent. In terms of material costs, there are much greater industry- and product-dependent differences (figure 6-12).

Typically, electronic purchasing processes are being implemented in three phases as shown in figure 6-13.

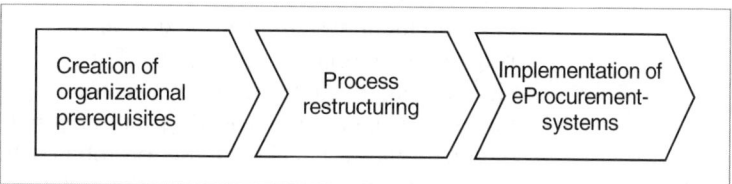

Figure 6-13:   Phase model of an eProcurement implementation

First, the procurement organization has to be "e-enabled" so that the requirements and responsibilities accommodate the new, standardized processes. Subsequently, the procurement processes must be restructured. Finally the eProcurement system can be implemented.

Once a business is eProcurement-enabled, it has access to different types of specialized procurement:

- Tenders
  Special requirements are usually procured through tenders.

- Catalogs
  Maintenance and repair goods are frequently procured via catalog systems.

- Auctions
  Perishable and used merchandise is almost always procured via auctions.

- Trade/Exchange
  Raw materials are generally procured through trade/exchange sites.

EProcurement increases efficiency. One mineral oil company, for example, reduced its procurement costs by 13 percent after implementing an international eProcurement platform. In bundling its purchases it cut the number of relevant suppliers and vendors from 120,000 to 6,000. Today, the eProcurement system handles 95 percent of all orders electronically. Other examples: A company of the chemical industry managed to lower its procurement costs by almost ten percent, while a leading credit card provider saved close to five percent.

Material purchase prices can also yield substantial savings. One automotive manufacturer, for instance, had already installed an eProcurement system. The next goal was to find more favorable procurement options for a part of the company's material purchases with the existing eProcurement system. After specifying the requirements, pre-selecting international suppliers, and conducting quality checks of the prototypes the company invited vendors to participate in an auction for third-party components. Given an order volume of approximately US$25 million, this corporation consequently saved close to eight percent in material costs.

6.2.2.7 Design-to-Cost

Design-to-cost could be defined as a product-oriented version of process optimization. In this case, product complexity is reduced with the goal of cutting material, production and development costs.

Product design optimization is contingent upon existing requirements as well as the simplest possible production method and/or assembly process (figure 6-14). The main driver here is the reduction of manufacturing components. The key element of the design-to-cost is either the design of a new or the redesign of an existing product so that it

- satisfies existing requirements with its attributes, benefits and quality,

- can be produced easily through minimization of costs, parts, and errors, consequently resulting in

- an assembly that requires minimum time and monetary investments and that delivers the highest levels of quality.

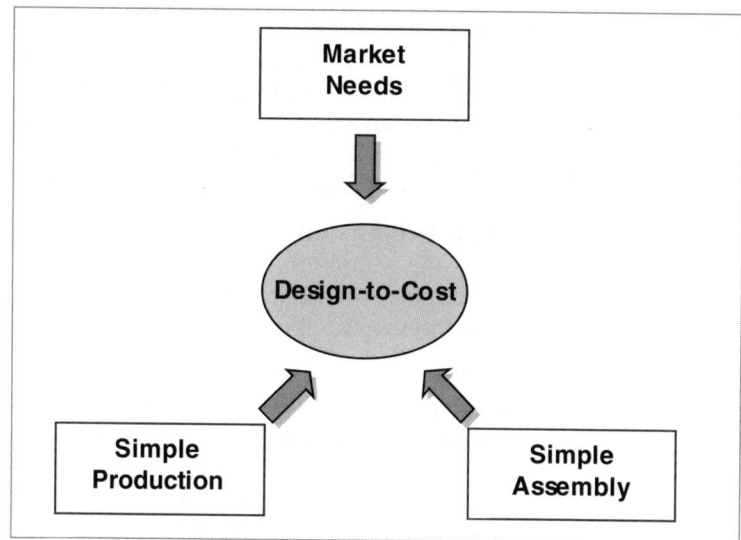

Figure 6-14:   Elements of product design optimization

Thus, design-to-cost offers measurable advantages in terms of cost as well as production time and quality, consequently affecting all segments of the value chain. It can be applied by most industries, for virtually all products and in most production stages. Improved product designs allow the reduction of variable and hidden costs. The latter stem from the restructuring/elimination of component functions that are no longer required. Simplified production processes may also render expensive special machines and equipment redundant. The design-to-cost process consists of three phases (figure 6-15):

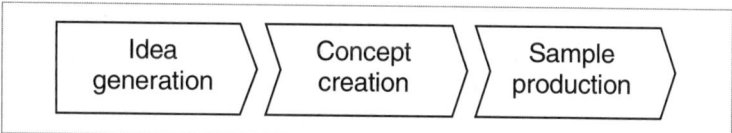

Figure 6-15:   Product design optimization phases

Phase 1   **Idea Generation**
In this first phase, as many improvement ideas as possible are collected and responsibilities for the potential improvement efforts are assigned. The ideas are neither evaluated nor ranked at the time of collection.

Concept Creation

In the second phase, all improvement ideas are assessed with respect to their technological feasibility and are then rated accordingly. Time and cost reduction potentials are estimated. For those ideas that receive good ratings, design blueprints are drawn up and potential suppliers are identified.

Sample Production

During the sample production phase, a first prototype is manufactured to demonstrate the functionality of the improved product and to validate the optimized production process.

A consumer goods company, for example, was able to reduce the number of parts necessary for the assembly of one product from 30 to eight thanks to optimized product design. An automotive manufacturer was able to realize cost savings of 45 percent while reducing the weight of the product by 66 percent by optimizing the design of one sub-assembly. A consumer electronics company minimized the complexity of a product to the extent that only 35 percent of the original assembly time was required after the improvements had been implemented. Swatch AG delivers yet another example of successful product design optimization: By reducing the components used in its products, it actually created a unique selling proposition.

6.2.2.8 Value Chain Optimization

One of the most important questions regarding business architecture optimization is that of the optimum degree of vertical integration: What part of the value chain should a company focus on and which parts should be outsourced?

Outsourcing refers to the commissioning part of the existing value creation process and therefore to the reduction of the value-added depth. The goal is to decrease costs while simultaneously enhancing the focus on the company's own core competencies. The outsourcing of parts of the manufacturing process and thus the purchase of these components from outside vendors can, for instance, yield reduced production and warehousing costs. Outsourcing does not have to remain restricted to production; it can actually encompass all operational areas. In general, outsourcing of part of the value creation process can make sense wherever costs may be saved without relinquishing strategic competitive advantages or where additional benefits, such as quality improvements, can be gained.

Consequently the outsourcing matrix makes outsourcing decisions contingent upon two dimensions: The strategic importance for the company and the current cost level of the part of the value chain in questions compared to that of external providers.

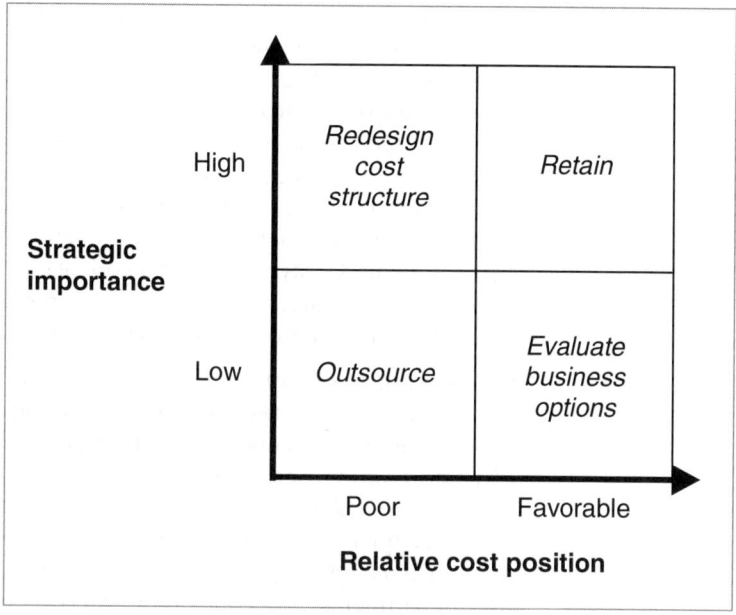

Figure 6-16: Outsourcing matrix facilitates decision making

The matrix yields four decision options (figure 6-16):

1. Functions that are of strategic importance to the company and that are being performed competitively should be maintained.

2. Functions that are of strategic importance to the company that are not being performed competitively should not be outsourced (long-term). Preferably, these functions should be cost-optimized.

3. Functions that are not of strategic importance to the company but are being performed competitively should be evaluated in more detail. In this case, the company could opt to outsource the function, offer it to external customers as a service or even develop a new strategic business unit (examples are Corporate Real Estate or IT).

4. Functions that are not strategically important for the company and that are not being performed competitively should be outsourced.

After Bosch GmbH had sold its former business segment "private networks" to KKR, Tenovis, the newly established company, was targeting a new strategic direction. The objective was to transform a conventional manufacturing company into a communication solutions and services provider. Services were to be offered for the company's own systems as well as those of other manufacturers. Given the changed corporate profile, production no longer played a strategic role for Tenovis. In compliance with the outsourcing matrix shown above, Tenovis made the logical decision to completely outsource its production (in September 2001, Tenovis owned only one production plant). To this effect, outsourcing had a direct impact on the company's profits – in the amount of more than US$5 million.

Outsourcing does not have to be restricted to parts of production. To increase its profitability, the oil company BP decided to focus on its core competencies. To that effect, BP outsourced all of its accounting operations to an external provider. Hence, the company was able to save 60 percent of its expenses in this area. The data quality also improved and the executive management were increasingly happy with the results produced. The accounting departments of competitor Amoco, who only recently merged with BP, and those of Arco, which was acquired shortly thereafter, were also successfully transferred to the same external provider.

NTL, a British telecommunications corporation has recently (March 2001) entered into an agreement with IBM, thereby outsourcing all of its IT services. The acquisition and integration of eleven companies was a gigantic challenge for NTL in the past two years. It was not always in a position to provide its customers with the desired level of IT services. The goal for IBM was to improve these services by consolidating all IT activities in Great Britain and Ireland. The agreement includes the transfer of almost 500 employees to IBM and is valued at more than US$2 billion. Ultimately, it is to yield cost savings in the amount of US$450 million.

6.2.2.9 Efficient Supply Chain Management (SCM)

For an efficient management of the entire supply chain, organization, processes, and systems must be aligned with corporate objectives. Moreover, in order to optimize the process, the supply chain must be integrated from the supplier's supplier to the customer's customer. Starting at one's own company, consistent multi-company processes must be implemented to reduce throughput times and consequently cut costs. Production systems and process optimizations frequently lead to an SCM program.

At the heart of the company-spanning supply chain optimization – which begins with the sourcing of raw materials and ends with shipments to customers – lies the implementation of compatible systems with a uniform data basis. Connecting the individual links of the supply chain and enabling direct data interchange can speed up processes. An additional advantage of this integration is the fact that in the event of supply chain interruptions (e.g., equipment failure) a contingency plan is immediately initiated alerting both ends of the supply chain. Efficient SCM is, however, not limited to purely logistical processes such as transportation, but encompasses all vendor and client relationships. This includes, for example, production activities and order processing. Efficient supply chain management supports the synchronization of demand and supply throughout the value chain and thus averts bottlenecks.

The supply chain management optimization procedure is divided into three phases: supply chain analysis, process design and systems implementation (figure 6-17):

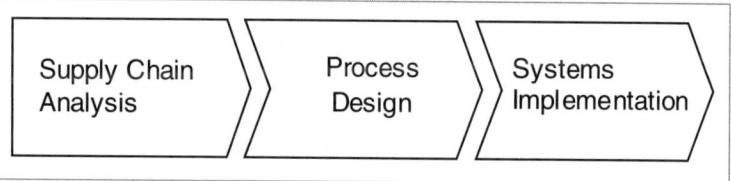

Figure 6-17:   Phases of efficient supply chain management

Phase 1   **Supply Chain Analysis**
This first phase analyzes the supply chain planning process. The entire logistics network, including all suppliers and customers, is identified and requirements are defined.

Process Design

The subsequent process design phase aims at mapping the logistics network requirements in a process and aligning all companies involved to this process. Suppliers and customers must be integrated into this process with no exceptions since the necessary organizational prerequisites (structures and processes) must be put in place at all involved entities. The optimization costs are usually shared by the companies on a prorated basis, as everyone involved benefits from efficiency increases.

Systems Implementation

In the third phase the actual SCM system is implemented by the companies who have joined the new logistics network.

A leading global food additives manufacturer serves as a prime example of efficient supply chain management. Due to declining profit margins and lower supply chain expenses among competitors worldwide, the company was compelled to optimize its supply chain. Manufacturing expenses were lowered while the order processing speed was increased. Inventory levels were reduced. The total savings came to approximately US$40 million, or close to 15 percent.

Cisco Systems also optimized its supply chain to improve its competitive position. While only four percent of all orders from vendors were handled online by Cisco in 1996, this figure now exceeds 90 percent. For about 75 percent of the products manufactured, the vendors actually handle the orders without Cisco's direct participation. These measures yielded an amazing US$600 million in savings, while inventories were reduced by 45 percent and throughput times accelerated by 25 percent. The level of customer satisfaction rose as well, because now 97 percent of all orders are shipped on time.

One of the world's largest mobile phone manufacturers was also forced to optimize/speed up its entire supply chain in order to compete successfully and commercialize product modifications more quickly. In addition, a more global orientation was to decrease costs even further. Supply chain optimization produced impressive improvements in this case, since the company attained a shipping accuracy of almost 100 percent while also reducing unproductive waiting periods by 85 percent and cutting finished goods inventories in some sectors by a factor of seven. The company also reduced the inventory range by three weeks in the process.

6.2.2.10 Systematic Customer Relationship Management

Customer Relationship Management (CRM) denotes consistent customer orientation, i.e., an alignment of all business processes towards the customer without compromise. This approach aims at attaining higher levels of effectiveness and efficiency in the areas of customer acquisition and relationship management while reducing costs through better knowledge of customer needs and wants. CRM is evolving into an increasingly critical competitive factor.

The successful introduction of a CRM system requires the restructuring of the organization and its processes as well as the implementation of the necessary IT systems. Customer-specific information must be collected, stored, and evaluated intelligently. The information generated allows targeted marketing to individual customers as well as customer groups.

CRM enables companies to identify and segment customers effectively. Moreover, customer relationships can be used profitably by linking existing data and information.

Another important aspect of the CRM system is the possibility to derive buyer behavior from the analysis of the collected customer data. This may lead to further optimization of business processes and, consequently, to additional cost savings.

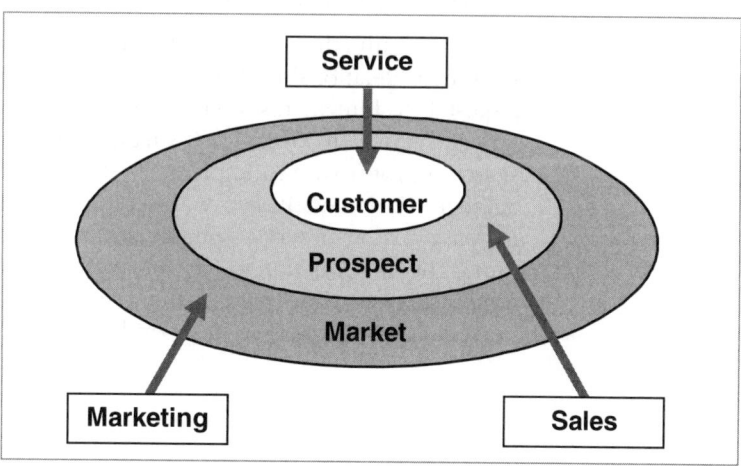

Figure 6-18:   CRM processes and target groups

The customer is at the heart of CRM (figure 6-18). To make a profit, businesses usually have to try to attract prospective customers' attention to their products through marketing, motivate them to buy, and bind existing customers by means of effective relationship management. A CRM system supports all of these processes efficiently by providing the necessary data for each of the tasks.

Efficient Customer Orientation Requires an Integrated System

Many companies have already taken their first steps toward CRM, e.g., through their Internet presence containing product information as well as through customer interaction centers (call centers), data mining processes or Intranets, all of which support of the distribution channels. Efficient customer orientation, however, is contingent upon the presence of an integrated overall system. This can also be independent software, which is part of an ERP system. In the context of such a CRM system, sales, marketing, and service teams as well as information systems need to be integrated and aligned with each other. Companies should, nonetheless, remember that the introduction of a new technology will not make up for potential shortcomings in a company's organization, processes or culture.

An Accenture industry analysis reveals that a large potential for profitability gains through CRM systems exists in the communications and high-tech industry (figure 6-19). Up to two-thirds of the profitability gap to the best-in-class may be eliminated by implementing integrated CRM systems.

One company that successfully increased profitability through Customer Relationship Management is Gevity HR. One of the largest U.S. human resource service providers, the company offers its services to over 8,500 smaller companies. Gevity HR was searching for a way to improve its data accuracy for better management decisions. It also wanted to automate its call centers to increase customer service quality and to reduce costs at the same time. The introduction of a CRM system led to improved customer service while cutting the duration of calls by about 25 percent to an average of 30 seconds.

An Internet bookseller improved the processing quality of customer inquires. The goal was to increase customer satisfaction through better interaction. Customer information was to be recorded intelligently to gain useful insight into consumer behavior. This company faced an additional problem: its employees were frustrated with the low level of efficiency of the existing system. Consequently, the personnel department found it rather challenging to fill job vacancies with qualified personnel. The introduction of a CRM system improved overall customer and employee satisfaction ratings considerably. The number of customer inquiries actually dropped twelve percent thanks to the improved quality of the information provided while employee motivation increased significantly.

	Electronics and high-tech	Communications	Chemicals
Top companies' ROS	33%	34%	30%
Average companies' ROS	13%	22%	16%
Total ROS potential	20%	12%	14%
CRM-related ROS-potential	13%	6%	7%
Share of ROS-potential due to CRM-systems	2/3	1/2	1/2

Figure 6-19:   Profitability improvement potential through CRM systems

6.3 Corporate Finance

The objective of corporate finance is to provide the capital required for business operations at the best possible price. Funding optimization can increase the value of a company considerably. There are two basic approaches: a reduction of the cost of capital in order to increase profitability and the improvement of access to capital as a basis for growth.

Freed up capital can either be used to repay borrowed capital or to fund continued growth. In most cases, the growth aspect is more important, and it will therefore be described in more detail. First, however, ways to decrease the cost of capital are presented.

6.3.1 Cost of Capital Reduction

Capital costs can be reduced through optimization of the capital structure and by way of efficient, centralized capital management. To identify potential areas of improvement, financial indicators should be generated on a regular and structured basis and then compared to competitor data.

6.3.1.1 Capital Structure Balance

If such benchmarking reveals that the weighted average cost of capital is higher than that for comparable companies, the funding sources should be reviewed. Often there are opportunities to adjust the debt-to-equity ratio, to replace expensive capital with cheaper funds, or to take advantage of better terms offered, e.g., on the international capital markets.

Debt to Equity Ratio

The starting point should be a review of the allocation of returns on investments between the debt- and equity capital sources. Profits are first distributed to creditors in order to satisfy the contractual interest payment obligations. The residual profits then yield the shareholders' return on equity capital. This value – and thus the shareholder value – increases as long as the return on additional investments exceeds the borrowed capital interest rate. This positive effect is called the *leverage-effect*. The return on equity can be raised through additional, relatively cheap borrowed capital.

This leverage effect calls for an increase of the debt-equity ratio and thus for increasing the share of borrowed capital to the maximum that the return on investments allows. As the debt level increases, this effect declines, however, since creditors demand higher risk premiums and only a few investments remain that can produce the required returns.

The separation of investment and financing decisions that is frequently practiced creates problems in this respect. It prevents decision makers from comparing investments with borrowed capital interest rates. Minimum returns on investment that exceed the average cost of debt capital should be stipulated for all investment projects.

Capital Sources In addition to changing the debt-equity ratio, capital costs can also be lowered through the diversification of a company's capital sources.

Figure 6-20 shows the correlation between liquidity and costs of capital procurement on the international capital markets along the example of Deutsche Telekom (largest German telecommunications provider):

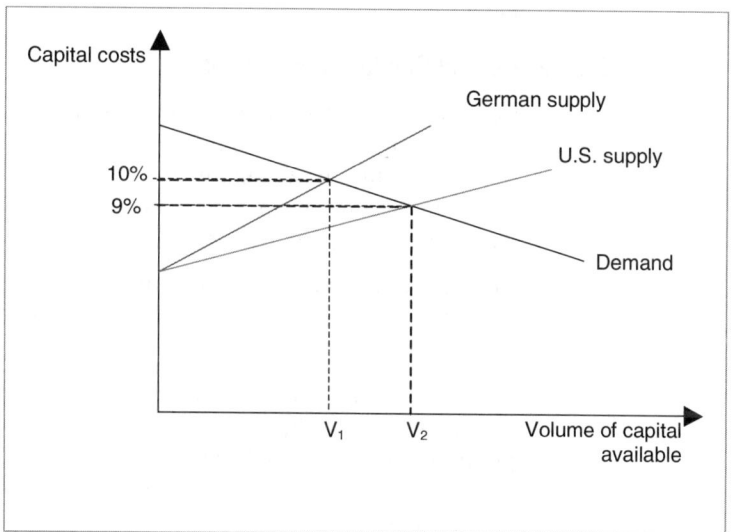

Figure 6-20: Capital costs in correlation to national capital market liquidity

Deutsche Telekom was faced with the choice between raising new capital on the German capital market (at ten percent average costs), or on the U.S. capital market (at nine percent for a higher volume). The difference in cost can be attributed to the higher volume of capital available on the American market. This example illustrates that companies should not only evaluate various domestic sources, but also review the funding options available on more liquid and therefore less expensive international capital markets when additional capital is required.

In addition to higher liquidity levels, international capital markets also offer alternative funding options. The *Eurobond markets* are a good example. Eurobond markets are not subject to government regulations, since loans are not issued in local currencies. There are no minimum reserve requirements for banks, thus they do not have to be included in interest rates to reflect opportunity costs. This allows companies to obtain loans at lower rates, which in turn lowers their average costs of debt capital. However, the risk on the Eurobond markets is higher given that the minimum reserves required in other markets do guarantee bank liquidity. Of course, currency risk must be taken into account for all international financing options, and needs to be hedged if necessary.

6.3.1.2 Central Capital Management

Central capital management can contribute to lowering the capital requirements of a company. It balances capital supply and demand within the company and can consequently minimize the financing volume from capital markets as well as the number of external funding transactions.

Internal capital allocations also allow global corporations to take advantage of taxation and/or interest rate differences between various countries.

Capital Transfer Group-wide tax payments can be optimized through the transfer of capital from a high tax country to a low tax country. And corporate tax rates vary substantially – from 16 percent in Ireland to 42 percent in Japan (January 2002; not adjusted for differences in the tax base).

Such capital transfers can be executed via dividends, patent and transfer price payments, or through *fronting loans*. In case of a fronting loan, one company issues a loan to another group company. A bank, with offices in both countries assumes the role of a mediator. From a legal standpoint, two autonomous contracts are being executed. One company invests a specific amount with the bank and the bank releases the same amount, deducting a small fee, to the other company in another country as a loan. The advantage of this construction lies in the fact that it bypasses capital flow restrictions while allowing the parties involved to take advantage of tax and interest rate benefits.

Procter & Gamble for instance realized substantial cost reductions upon implementing a global capital management system. Until the mid 90's, all international Procter & Gamble subsidiaries operated their own financial divisions, which made all relevant decisions regarding investment policies, financing, cash management, and currency risk management. These activities were coordinated on an international basis through the group's headquarters, which issued target limits, such as minimum returns on investments. Given that the conglomerate's production activities were centralized to a large extent, the individual national organizations were tightly interwoven through a multitude of financial transactions. International transfer payments were subject to currency risks. The introduction of a global treasury management yielded annual savings in eight-figure amounts. Bypassing banks and bundling transactions between intra-country organizations reduced transaction fees. The introduction of a central management saved additional costs in the area of currency risk hedging. Overall, capital costs dropped significantly because a larger volume of internal loans was granted through the company's central capital management.

6.3.2   Funding Growth

Companies have a choice of funding their growth either internally by utilizing capital more efficiently or externally by raising additional equity or debt. A number of specific options are available in both cases.

6.3.2.1 Funding Growth with Internal Capital

The measures described in this segment free internal capital. While these funds can be used to decrease debt and thus capital expenses, they can also fund continued growth.

Liquidation of Assets that are Not Being Utilized

The liquidation of non-utilized assets frees up capital without adverse effects on ongoing business operations. For instance, many traditional companies own a lot of real estate, which is not being sufficiently utilized. Deutsche Post AG, the former German postal service monopoly, liquidated unused assets to grow the core business. After an analysis of the total real estate holdings and a subsequent optimization effort, unproductive property was sold. The company reinvested the sales proceeds – on the order of billions of euros – for the acquisition of more than 20 European parcel services. Deutsche Post thus became Deutsche Post World Net.

Sale-and-Lease-Back

The *sale-and-lease-back* method is primarily used to fund rapid growth. In this case, portions of the organization's fixed assets are sold and then immediately leased for continued use. The leasing expenses, however, are usually higher than those of the fixed capital assets and depreciation. Because this process does have a negative impact on profitability it is most appropriate for businesses whose growth requires a large amount of capital.

Cutting Payment Terms on Receivables

The intermediate financing of receivables often also binds substantial amounts of capital. Tightening payment terms and monitoring compliance with these terms can therefore release capital for the funding of growth. However, this action puts pressure on customers, which could have an adverse effect on customer relations.

Factoring

Similar to the shortening of payment cycles, *factoring* allows businesses to reduce the amount of capital bound in receivables. In this case, a company's receivables are sold even before they become due to improve liquidity. Given that receivables can only be sold at a discount in this case, factoring incurs increased transaction costs.

Inventory Reduction

Inventory reductions decrease the funds tied down in current assets and thus the need for capital. Improved logistics, which can be realized for example through Supply Chain Management and more efficient production processes, releases a substantial amount of capital in this area.

Box 9:   How Effective is the Reduction of Capital Employed?

In considering measures that lower the amount of capital employed it should be noted that they only have an indirect impact on costs, in particular in terms of profitability. The EVA equation reemphasizes this fact. The EVA is calculated by deducting the capital cost from the after tax profits.

Here, the cost of capital equals the product of the weighted average cost of capital and the capital employed. If the capital employed is reduced by ten percent at a capital cost rate of ten percent, the EVA is improved by one percent accordingly. The same effect can, however, also be achieved by directly increasing the after sales profits by one percent. Therefore companies should be aware of the fact that increasing a business' value through a reduction in capital employed requires a far bigger relative effort than measures aimed at increasing after tax profits.

6.3.2.2 Funding Growth with External Capital

International capital markets should always be considered when capital is to be raised. As discussed in the context of capital cost minimization, these markets offer a larger total capital volume.

In conjunction with its public offering, Deutsche Telekom approached various capital markets simultaneously to obtain the funds required for investments. The European telecommunications deregulation catapulted Deutsche Telekom into international competition. To defend its competitive standing, Deutsche Telekom invested massively into new infrastructure, such as its mobile phone network. However, the company initially, encountered substantial problems in procuring capital.

Deutsche Telekom no longer had access to government resources, since the Maastricht treaty restricts the options for household deficits. The German capital markets were not sufficiently liquid. Exclusive placement on the German stock market appeared to be impossible given that Deutsche Telekom's value was estimated at approximately US$30 billion, which exceeded all other German stock listings. Plus, up until a few years ago, Germans were hesitant to invest in stocks. In 1996 only five percent of all German citizens actually owned stock. In America, about 25 percent of the population held shares; the British came in at 20 percent. In the mid 90's Deutsche Telekom also had to compete with many other former government-held companies in Europe, who were trying to attract a comparatively small number of potential investors. The only way Deutsche Telekom could have obtained the required capital in Germany would have been by promising extremely high dividends. This was, however, contradictory to the objective of becoming profitable.

Consequently, Deutsche Telekom executed its privatization in several stages and offered its stocks simultaneously in Frankfurt, London, New York, and Tokyo. In a first step, 25 percent of the company valued at US$6.5 billion, was offered on the stock exchanges. At that time, it was the largest IPO in Europe and the second largest worldwide.

6.3.2.3 Funding External Growth

While cash is required for organic growth, stocks can be used as an alternative currency in mergers and acquisitions. The financial transaction can be executed in cash, through a stock swap, or as a combination of both.

This choice influences the shareholder value of the new larger company as well as the type and scope of risk that has to be borne by shareholders at both ends. In principle, acquisitions should always yield synergy values based on the companies' joint business activities and should translate into higher stock market valuations. This value is divided between the stock price premium of the company to be acquired and the buyer's anticipated net profits.

The portion allocated to the stock price is disbursed to the shareholders of the acquired company. Consequently it directly affects the future profit potential of the new company. This allocation is termed *Goodwill* and refers to the difference between the purchase price and the actual book value of the acquired company. The higher it is the more financial pressure it produces in terms of future depreciations. In the United States, companies have the option to show only the book value of the acquired company in the balance sheet by applying the *Pooling of Interest* method. At this time, discussions concerning the abolishment of this practice are, however, underway. Depreciations have a direct adverse effect on the earnings per share, while the *EBITDA* remains intact due to the deductions of goodwill amortizations (the "A").

The remaining net profits are allocated to the shareholders of the new company. To this effect it is of particular relevance whether the acquisition was a cash or stock transaction, since the payment method determines the future shareholder structure of the new company. Therefore, only these two options are further discussed: Cash acquisition and stock trade.

| Cash Acquisition | When cash is used to acquire companies, the shareholders of the buyer are the sole owners of the new larger corporation. This method requires them to bear the full risk if the anticipated synergy value is not realized (figure 6-21) and a net loss is suffered instead. The buyer's shareholders are often therefore only willing to pay a small price premium over the current market valuation. |

A special version of cash acquisitions is the *Leveraged Buyout* (LBO). It was especially popular during the 80's and is currently regaining ground due to the weak capital markets. In case of a leveraged buyout, a company is purchased by a financial investor or by its own executive management (Management Buyout, MBO) utilizing borrowed capital. The resulting high debts force the company to take advantage of existing cost reduction potentials and to dispossess of less profitable business divisions. If the streamlined company is subsequently offered to the capital market it usually attains a higher valuation because of its increased profitability. To motivate shareholders to approve a leveraged buyout, the financial investor often offers a premium over the current stock price. If several competing investors are involved, they are frequently willing to pass on a large portion of the anticipated synergy effects to the shareholders via premiums. For the financial investor, the risk of an unsuccessful acquisition is thus increased, given that the investor is solely responsible for the realization of the synergy effects.

| Stock Trade | When companies are acquired through stock trades, the buyer runs the risk that once the merger is complete the shareholders of the acquired company may hold a majority stake in the new corporation (figure 6-21). As a tradeoff the new shareholders bear the pro-rated risk if the anticipated synergy value should not materialize. In case of stock trade acquisitions it must also be determined whether the trade is to be transacted based on a fixed number of shares or based on a set company value. |

If the trade is based on a fixed number of shares, the transaction value may change from the date the offer is disclosed to the date the contract is closed. These price fluctuations impact the shareholders of both companies, but they have no influence on the new company's ownership ratio. For the buyer's shareholders such transactions entail the risk that the market valuation of the acquired company declines. The acquiring company also runs the risk of the deal failing if its share price declines and the shareholders of the company to be acquired reject the swap offer.

153

			Advantages	Disadvantages
Cash		Buyer	100% ownership	Risk that synergy effects will not materialize
		Seller	Fixed premium over stock price	Abandonment of owner-ship
Stocks	Fixed number of shares	Buyer	Fixed stake in new com-pany	Risk that synergy effects will not materialize, rejection of acquisition offer in case of target company stock price decline
		Seller	Possible premium over stock price	Buyer stock price may decline, risk that synergy effects will not materialize
	Fixed value	Buyer	Fixed purchase price	Possible loss of share-holder majority, price risk for own shares
		Seller	Fixed premium over stock prices	Risk that synergy effects will not materialize

Figure 6-21:   Advantages and disadvantages of purchasing options

Yet another method of structuring the stock swap consist of issuing shares with a fixed total value. Since the number of shares allocated to shareholders of the target company is not determined before the contract is ratified, the ownership ratios are then contingent upon the price level of both shares at the time of closing. The buyer's shareholders bear the full price risk for their own shares if this trading model is applied.

One example of a stock acquisition was the 2001 takeover of Voicestream Wireless by Deutsche Telekom. Voicestream's shareholders were given the option to go for a straightforward stock trade or for a combined approach. The cash share of this US$28 billion acquisition was restricted to US$4.23 billion. Consequently, a fixed price of 3.2 Deutsche Telekom shares plus a cash component of US$30 were established in July 2000. The take-over offer included a 60 percent premium over Voicestream's stock price. Voicestream shareholders were awarded substantial protective rights. They had the option to rescind from the take-over if the Deutsche Telekom stock price were to drop below US$33. This had a negative impact on the valuation of Deutsche Telekom's stocks. On the day of the announcement, its price fell ten percent to below the US$50 mark.

A high market valuation of the acquiring company favors stock swaps over cash transactions because in this case, the shares of the buyer can be employed as an acquisition currency.

If it is likely that a company's share price declines prior to the closing of a stock swap-based acquisition contract, it would be preferable to choose the *fixed price offer*. This way, the buyer can demonstrate confidence regarding the anticipated synergy effects and burdens part of the risk. This may actually stabilize the stock price since the capital market is relieved of some of the uncertainties connected to the acquisition's success. If the buying company has access to adequate liquidity or the option to raise debt capital, all the merger profits would go to its shareholders in case of a *cash offer*.

In the past stock trading transactions have gained considerable ground and are now making up about 50 percent of all larger acquisitions. By 1998, pure cash acquisitions, on the other hand, had dropped from a level of sixty percent in the late 80's to a mere 17 percent. Studies reveal that the shareholders of acquiring companies have fared significantly worse in case of stock swaps compared to cash transactions in the period after the deal announcement. Companies are increasingly willing to execute acquisitions even if they do not have the liquid funds to proceed. They relinquish the realization of short-term shareholder value but strengthen the new company's market position with longer-term synergy potentials. Consequently, higher company valuations are feasible in the long-term.

Chapter 6.3 has shown that the optimization of corporate financing may contribute significantly to value creation. It has a direct impact on both value creation levers by increasing profitability on the one hand and creating a foundation for growth on the other.

6.4 Human Capital

The importance of employee knowledge, skills, and motivation is often emphasized, but is very hard to gauge, which leads to an awareness problem. When companies are under cost pressure or face a crisis, they frequently make the mistake of underestimating the importance of *human capital.*

6.4.1 Personnel – More Than Just a Cost Factor

Cisco has demonstrated impressively how valuable excellently trained personnel can be for a company. In the mid 90's during the early expansion phase of the Internet, Cisco was embarking on a period of rapid growth, which could only be realized through acquisitions. Instead of evaluating potential candidates based on financial concepts alone, Cisco selected the companies it acquired mainly because of their ability to come up with new ideas: Cisco rated take-over candidates almost exclusively based on the quality of their development engineers. Allegedly, Cisco credited these companies' values with a lump sum of up to US$2 million per engineer. From 1994 to 2000, Cisco considerably increased its company value via the take-over and fast integration of more than 70 companies and several thousand engineers.

In dynamic markets, competitive advantages over other companies usually do not last very long. What makes or breaks companies is their ability to learn faster than others and to adapt to the changing market situations. Consequently, the institutional and individual learning processes are essential. Both processes are being impacted by the corporate culture on the one hand and by the abilities of the employees on the other. Yet these are also interdependent: a company's culture influences what types of people it can attract and how they function and develop inside the organization, while the skills and personalities of the employees in turn affect a business' corporate culture.

The long-term success of businesses hinges on the quality of their people. Nonetheless, highly qualified employees tend to change jobs more frequently and largely independent of economic cycles. Companies are therefore continually at risk of losing competitive advantages they have gained in terms of human capital. Hence it is imperative to find a balance between the interests of the company and those of the employee.

In addition to an adequate incentive system, the personnel management strategies – which range from recruitment to separation – have a decisive impact on whether a company is going to be considered a good place to work or not. The individual personnel management processes are shown in figure 6-22.

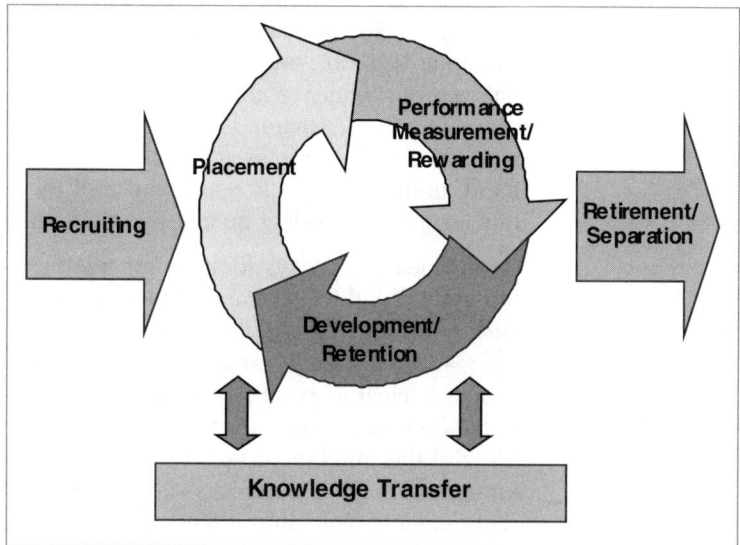

Figure 6-22:   Stages of an employment relationship

6.4.2   Human Resource Management Phases

To optimize the contribution of human capital to the company's success, all phases of the employment relationship – beginning with the selection of suitable candidates and ending with the termination of the employment – must be managed systematically. The following sections therefore explain the elements of a successful personnel management program. The different phases of the employment relationship are described in detail and illustrated with practical company examples.

6.4.2.1 Recruiting Employees

Companies or business units post jobs and, according to the objectives, stipulate specific skills and abilities. If growth is the goal, the company needs creative and entrepreneurial types to reach it. Companies that are aiming at consolidation will find efficient implementers more suitable for their needs. Hierarchical depth issues also determine the profiles of potential employees.

To find a perfect match for their needs, companies communicate an image to the job market that is consistent with their corporate strategy. This image contains general values as well as the company vision. It also explains what is expected of future employees and reflects upon career opportunities and compensation packages. The Media & Entertainment giant Bertelsmann, for example, bets on young team players who take on entrepreneurial responsibilities within the organization. In its job posting campaign, Bertelsmann is consequently announcing "Entrepreneurs Wanted." German automotive manufacturers, on the other hand, utilize an equally successful tool by leveraging their positive image and product quality reputation to attract new recruits.

Companies are always looking for people for specific positions, but they should also create talent pools from which they can later select suitable employees for future job openings. If a business is looking for a candidate to fill a specific position, the job profile is clearly specified in the advertisements. If, on the other hand, new employees are to be hired for a talent pool, their profile and the number of persons to be hired will be based on an integrated corporate career strategy rather than on specific job requirements. Generally companies decide to utilize one or the other recruitment strategy. Procter & Gamble, for example, develops its executive management from within the organization and recruits managers from an internal pool, unlike General Electric who looks for external candidates that meet a specific profile for all open positions, including those on the executive management level.

As employee qualification profiles become more demanding, the selection of suitable candidates gets more difficult. Successful corporate recruiters use a wide array of media in addition to conventional job ads to create a maximum pool of candidates that might be of interest to the company.

Internet

Companies that utilize the Internet as an employee recruitment medium, adapt the recruitment process, the technical infrastructure and the application processing procedures to the requirements of this new channel. Online application forms are designed so that their completion by the applicant delivers sufficiently meaningful information for a preliminary assessment. Response times to application mails are reduced to the point that candidates are not lost to competitors that might react more quickly to applications that have been sent simultaneously.

Information Campaigns

Companies also have the opportunity to interact with potential candidates through information campaigns, university events, job fairs, and industry conventions. University cooperation offers long-term opportunities. They allow companies to get in touch with potential candidates at an early stage in their development and create loyalties. DaimlerChrysler, Samsung, Siemens, and Nokia are renowned for their policies of establishing direct university contacts with young academics via scholarships, exchange programs, internships, and corporate presentations. For promising candidates, the orientation and application process frequently begins while they are still in school.

Internships

Internships are offered to students who have successfully completed a demanding selection process. These students have the opportunity to participate in a thoroughly structured program parallel to their studies, which encompasses practical on-the-job-training and extracurricular seminars. Microsoft, Ford, Xerox, and IBM, for example, have obtained direct access to highly qualified candidates through cooperation with leading universities, such as the MIT with its Center for Information Systems Research (CISR) and the Center for Innovation in Product Development (CPID). Cisco Systems is frequently present at career fairs, at universities, and conventions of related industries, and it has consequently registered 500,000 applicants in its recruitment database.

To recruit the best applicants, companies must define meaningful and efficient selection processes. To this effect, the adequate selection instruments that apply to a specific open position must be determined. Given that candidates usually send applications to several potential employers, prompt reactions are the key to successful recruitment processes. Nonetheless, this must not be done at the expenses of candidate quality. Companies can also motivate candidates to make prompt decisions if they offer very attractive jobs for a limited time period. This may prevent applicants from waiting for other opportunities.

Even in times of economic downturn, employers should continue recruiting and evaluation efforts. It is important to remember that someone who may not be hired today might become a wanted candidate tomorrow. For this reason it pays off for companies to create mutually beneficial relationships with applicants. One possible instrument for flexible personnel planning is a bonus system that offers accepted applicants a compensation for deferred hiring dates.

6.4.2.2 Staffing Positions and Employee Reassignment

Employees deliver peak performances if their task profile matches their individual skills, know-how and interests. In this context, the assigned tasks should be challenging – and in the sense of continued development – inspiring. Only this matching ensures that employees are motivated to commit and expand their full potential.

Many businesses have explicit or at least implicit career path definitions in place. Rotation mechanisms allow employees to change their positions and sometimes even the countries they work in to expand their experiences and hone their skills. Their scope of responsibility increases simultaneously – in a step-by-step process. Gillette fosters frequent job changes within the home country and in other countries. Half of the employees, who work outside of the home country, are already working in their fourth country. Foreign assignments are not considered dead-ends but are honored as part of the career development. SAP also fosters job exchanges. Its executives complete an average of three to four international assignments.

Career development programs should balance the needs of the company with the expectations of the employee. Consequently, job policies, company-wide workforce planning and general career strategies must be coordinated. Otherwise, reassignments will not be smooth and the interests of company and employee will not be taken into account sufficiently.

6.4.2.3 Measuring and Rewarding Performance

To align employee performance directly with the company goals, the individual's targets must be derived from them. This can be achieved through a cascade-style top-down goal setting. Given that companies usually have more than one goal, the institutional goals should also be anchored to the individual goals of the employees.

Balanced Scorecard

For management positions, the use of balanced scorecards has proven effective. It measures the level of attainment of a company's strategic goals by translating visions and company objectives into concrete actions and metrics. The balanced scorecard program communicates the company goals and allows managers to actively impact business results. It usually evaluates performance based on four perspectives: Finances, products, customers and employees (figure 6-23).

Non-managerial staff members often do not have an impact on all relevant perspectives. Nonetheless, their performance evaluations should also be based on several criteria. Their typical layout typically deviates considerably from classic balanced scorecards. Obviously, the evaluation criteria depend on the type of industry as well, given that different abilities are crucial for every company's market success. For instance, management consult-

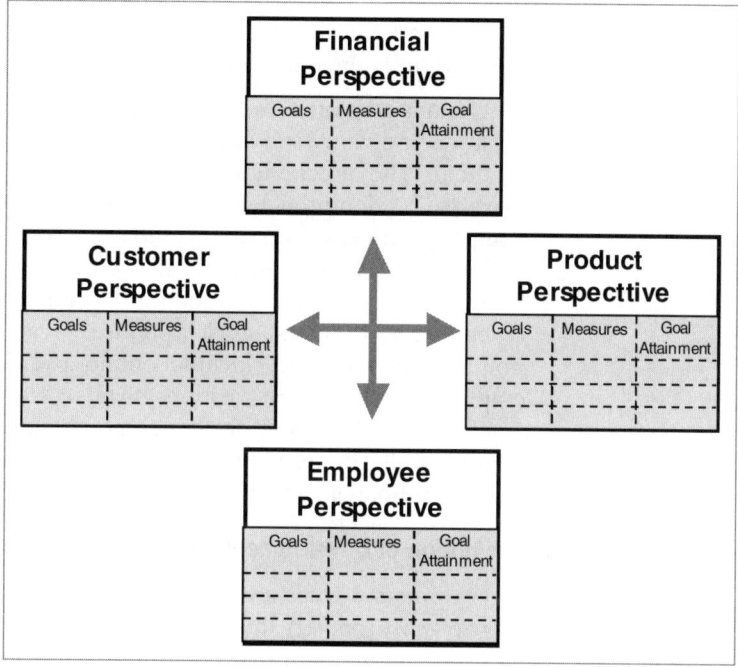

Figure 6-23:   Scheme of a balanced scorecard

ants are frequently evaluated based on skills in the areas of problem solution, teamwork, communication skills, personal initiative, and client relations management.

While companies expect their employees to perform adequately, employees expect fair compensations for this performance. But fair rewards are only possible if the level of performance is transparent to the employee and to the company. The employer must therefore clearly define suitable metrics, instruments and intervals for such performance reviews, and communicate this information comprehensibly. Six to eight metrics are recommendable in this context. The degree to which each target metric should be reached within a given interval must be agreed upon. Actual individual employee and department performance is then measured based on these defined target metrics and compared to the expected target attainment levels. The more transparent these performance indicators are, the easier it is for employees to raise their performance based on the performance criteria. General Electric is known for the clearly defined and comprehensible quantitative and qualitative evaluation factors it applies to areas that can be directly impacted by the respective employee. These factors have been aligned with the overall company targets and those of individual divisions.

To commit employees to the attainment of goals, the relationship between individual performance and rewards is communicated to them. The company's decision about whether to focus on financial or non-monetary rewards must be consistent with the company strategy and the company's image, and it is made based upon market analysis and employee surveys. Financial incentives may be issues in the form of performance-based salary components, bonuses, stock options, or capital stakes. Possible non-monetary rewards include public awards or entrusting the employee with more responsibilities. Hewlett Packard and Nokia, for instance, use non-monetary rewards to a much greater degree, while Bertelsmann, SAP, PepsiCo and Disney place a lot of trust in financial compensation.

6.4.2.4 Promotion of Development and Continued Education

Promoting Development Through Career Management

Developmental goals are discussed and established with the employee early on. To this effect, employee ambitions and company needs are harmonized. These developments can be fostered both by transferring responsibilities to the employee and through formal training. In this context, both sides are required to optimize the employee's performance in the interest of the company.

Bertelsmann, Johnson & Johnson, and Nokia develop and challenge their young managers by assigning responsibilities to them after only two or three years with the company. Employees thus have the opportunity to prove their true potential in a clearly defined area of responsibility. Moreover, the leeway granted motivates the staff member and allows him or her to grow with the assigned tasks.

The results of regular performance reviews provide the basis for the employee's career development. GE has been successful with its annual "Session C" performance reviews. The program keeps managers abreast of their employees' strengths and weaknesses. In connection with performance reviews, coaching or mentoring programs have proven helpful. They provide the employee with trust-based career guidance and orientation from an experienced co-worker. The employee discusses the results of the performance review with the coach in an evaluation interview. In this meeting, employee and coach jointly identify the strengths and weaknesses of the employee and subsequently derive the required consequences, such as for example training needs or foreign assignments.

Many companies actually have an explicit career management function, which is available to those with specific questions about their career development. The career management function defines and organizes development and performance support programs for different groups of employees. It is therefore in control of skill-building activities in coordination with the work requirements. Specific training curricula are assigned to each career path for which a training plan is developed in cooperation with internal and external providers. The company then informs all employee groups about mandatory training courses and also recommends specific courses beyond these requirements. Training budgets are often substantial. GE, for example, invests US$1 billion per year in human resource development.

6.4.2.5 Employee Retention

Personnel Fluctuation is Very Expensive

High levels of fluctuation cause enormous expenses. Building the loyalty of a company's best performers is therefore one of the highest management priorities. For Jack Welch, the former CEO of General Electric, this topic was at the very top of the agenda. He considered it his personal responsibility to find the best employees for GE and to retain them. Cisco is another excellent example for a company dedicated to personnel retention. In 2000, its 8 percent fluctuation rate was far below the 20 percent average of its competitors.

For employees to be productive, they have to like their work environment and feel confident that their company will meet their expectations. Financial rewards, attractive career options and a positive working environment enable companies to build high material and psychological exit barriers. Especially the balance between work and private life plays a crucial role in this context. Work hour accounts, part time job opportunities, and other flexible work time options have proven effective. Fostering a spirit of belonging through company funded events and team leisure time activities (which could be linked to project milestones) can motivate especially younger employees and create a sense of loyalty. The overall reward system – in particular in terms of balanced monetary and non-monetary rewards – depends largely on the cultural area.

If a company's culture creates a sense of "we" in its staff, the company is transformed from an anonymous workplace into a social environment in the eyes of the employees. This kind of "we" attitude can, for example, be fostered through sports competitions against teams from other companies. In the United States, these "Corporate Sport Challenges" between major corporations are even broadcast via television.

Symbolic sports competitions can also have a direct impact on increased performance. A worldwide logistics company, for example, used an intra-company competition to solve manual package shipping problems. The tournament was based on the professional baseball World Series. Teams were assigned points on a daily basis when different criteria were met. The weekly champion from every location advanced to the next round where he or she competed against teams from other locations. This exercise did not only foster a spirit of sticking together within the teams, but also yielded significant improvements over earlier results in all performance rating categories. These high quality standards were successfully maintained beyond the end of the tournament. In addition to integrating its employees, the company consequently also directly created value.

Companies should offer their employees opportunities to provide feedback on areas of concern or problems that are responsible for their dissatisfaction via surveys and coaching programs. This aids in preventing resignations of highly qualified employees if their dissatisfaction is identified and their input received in due time. While many executives have misgivings about institutionalized feedback from those who report to them, companies who have implemented 360-degree ratings have seen extremely positive results.

6.4.2.6 Knowledge Management

As long as the knowledge of each and every employee is not interchanged with other employees or stored in databases, it is just as elusive as the employees. In times of high fluctuations, companies are therefore well advised to be prudent in ensuring that the knowledge and know-how acquired inside the business is not lost along with leaving employees.

Effective Knowledge Management

Moreover, effective knowledge management aims at enabling employees to share in the knowledge of all of the company's employees in order to raise performance standards. To this effect, data must be accessible in such a way that required information can be found easily and that employees are not swamped with irrelevant facts. Figure 6-24 shows suitable methods for institutionalizing a knowledge-exchange in companies.

✓ The company generates a database as a virtual library, in which knowledge is stored and made accessible to the employees.

✓ One department inside the company administers this database.

✓ The company introduces uniform processes for storing and accessing information, in cooperation with the employees.

✓ Accessibility and systemization information is communicated to employees on an ongoing basis.

✓ The company predefines standards and templates that allow employees to identify and collect knowledge.

✓ Each employee's know-how is electronically recorded and integrated into units and projects of a knowledge database.

✓ The responsible department prepares the knowledge in terms of content and format and systemizes it based on a structure that facilitates ease of use.

✓ Special areas of knowledge are indexed accordingly to allow relevant employee groups to obtain and access this information efficiently (channeling of knowledge exchange).

Figure 6-24:   Aspects of implementing a central knowledge management system

In internal company forums employees have the opportunity to exchange job-specific knowledge. This type of experience and knowledge sharing can be done online, but also at regular meetings. With the help of access statistics, surveys, and evaluations by employees, the knowledge exchange and database generation processes are continually improved. All relevant external information and news should also be made available to the employees.

6.4.2.7 Termination Procedures

Termination of one or several employees can have a profoundly negative impact on the atmosphere within a company. It could also be understood as a sign of economic problems. Alternatives to layoffs are wage increase waivers, a shorter workweek, long-term vacations with partial compensation, or job sharing. If job cuts cannot be prevented, fair treatment of those concerned and honest communications are of paramount importance. It is particularly crucial to assist terminated employees in finding new jobs. In its relationship with the remaining employees, the company must clearly communicate why the lay offs were necessary. It must also address potential job worries of the remaining staff.

During times of economic boom, companies must be especially prepared to lose employees for personal or professional reasons. In this context, processes must be institutionalized that ensure continuous coverage of the functions affected and that enable knowledge transfers. Exit interviews should become an elementary part of the termination process. These meetings allow employees to state the reasons for their resignations frankly and without concerns about their future career. This kind of frankness frequently provides companies with valuable information on internal improvement potential in the area of human resource management.

Exiting employees can be of further benefit to the companies if both parties stay in touch after separation. Relationships with former employees provide an additional channel to new customers or new potential employees. These contacts can be institutionalized via alumni networks. All management consultants place great emphasis on maintaining such contacts, given that former employees frequently move to influential positions in client companies. McKinsey and Company have one of the best alumni networks, which offers all alumni access to databases and fosters regular meetings. This allows the company to stay in interactive contact with former employees.

6.4.3   Analysis and Elimination of Performance Deficits

Optimized human resources management aims at maximizing the human capital value contribution. Its individual phases are particularly helpful in uncovering and eliminating *performance deficits*.

If performance deficits are detected at entire departments during performance evaluations, this could point to deficient incentive systems, inappropriate transfer prices, or individual department head deficits. The true cause can be established when similar internal departments are compared and interviews conducted. If organizational provisions are responsible for failures to reach targets, the incentive system should be adjusted as described in chapter 5.

If individuals fail to attain goals – whether they are managers or employees on lower levels – the cause could be two-fold: ability and willingness of the employee. Staff members must, after all, be motivated and able to reach goals. Employees may fail to reach goals and meet expectations because they:

- do not know what is expected of them,
- do not have the required authority,
- have been provided with insufficient resources,
- do not receive any feedback on their performance,
- are punished for good performance,
- are rewarded for inadequate performance,
- are ignored no matter whether they perform well or inadequately or
- do not possess the required skills.

Employee Responsibility Restricted to Inadequate Skills

Of all of these causes, the employee is only responsible for one: inadequate skills. In practice, however, the other aspects are, far more often responsible for the failure to attain goals – and they are the responsibilities of management. Unfortunately this is frequently overlooked, either intentionally or unintentionally. Instead, training or employee replacements are demanded, while the true causes are not analyzed in detail.

Incorrect motivational structures appear often in companies. A manager, who stayed below his/her annual budget because of good management, is punished by superiors when the budget is reduced in the following year. Another common mistake is to entrust employees who have performed excellently on an unpopular task permanently with this type of work. It is not difficult to figure out that these types of negative motivation will not yield the desired results.

Inappropriate rewards do not just harm the employees, but are also bad for the company's customers. The service department employees of an equipment manufacturing company were asked by their superiors to perform preventative maintenance in addition to the repairs they were conducting on machines at customer's facilities. The performance in this case was measured based on the time required for each service job. Superiors used this metric in rating the employees' performance and determined compensation accordingly. In other words, employees who spent more time than absolutely necessary with the client were punished. In consequence, the maintenance work was neglected.

If there are performance deficits, management should first verify if it is responsible itself for the failure of individual employees to meet targets. Resource allocations, competencies, and especially incentive structures must be carefully reviewed and aligned correctly. If the problem is really with the employee, an assessment of his/her potential can help in establishing whether the employee can obtain the required qualifications through training. If this is not the case, disciplinary consequences, including reassignment to a more suitable position or termination, are unavoidable.

Continuous Analysis of Failures to Attain Goals

The elimination of performance deficits is never a one-time initiative. On the one hand, companies continuously hire new personnel, while other employees are assigned to different responsibilities or promoted. On the other hand, goals and target metrics of individual departments and employees also change due to market developments and internal restructuring. Consequently, the causes of failures to meet expectations must be analyzed regularly after each performance review. This is absolutely essential if reasonable organizational regulations are to support the attainment of goals at all times and if employees are not just to have the capacity to meet expectations, but are also motivated to do so.

6.5   Investor Communication

Investor communication aims at the increase and long-term maintenance of a company's market valuation through the provision of information on planned growth and profitability increases. To achieve this, it must drive investor expectations in a positive direction.

To this effect, the right recipients must receive relevant information. Communications instruments must be used skillfully to attain sustained value creation on the capital market.

6.5.1   Investor Awareness – The Key to Valuation

The value of a company is made up of the weighted sum of investors' future expectation scenarios. Consequently, the successful employment of the value creation levers must be communicated to investors to affect their anticipations positively. Failure to do so can cause actual value increases to go unnoticed in the market.

Derivation of
Development
Scenarios

Investors derive various development scenarios from the information communicated by the company. They take the market environment into account, along with the historic value curve of the company and their own individual expectations. These scenarios describe possible future developments in terms of revenues and profitability. Investors tend to deduct a risk factor from companies' revenue and profit projections, so that external expectations are usually more conservative than those officially communicated by the company. It is nonetheless possible, that outsiders expect a scenario that exceeds business plan projections.

Different investors usually have different expectations. Figure 6-25 shows the probability curves of these different scenarios. Each scenario is assigned a probability level that is based on the relative number of investors who believe this expectation to be realistic. As a rule, the sum of investor expectations yields a normal distribution, in which scenario Y stands for a company development with faster growth and higher profits than scenario X. To this effect, scenario Y could reflect a business plan based development, while scenario X may include a safety deduction.

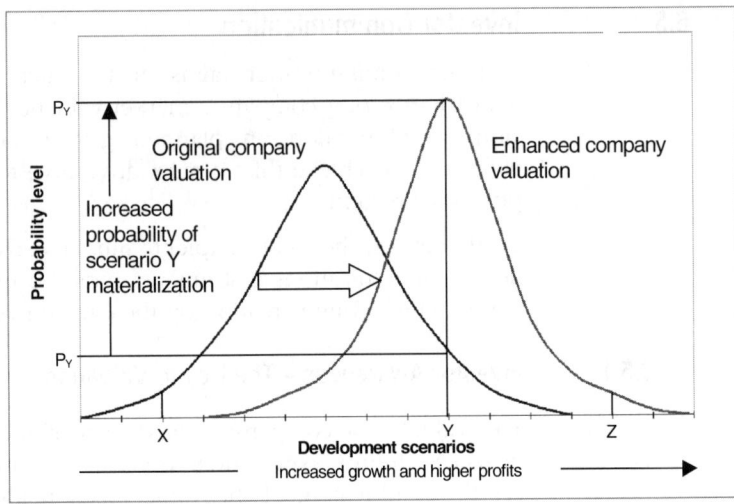

Figure 6-25: Probabilities of company development scenarios

Improved investor awareness and increased investor confidence could consequently have a substantially positive impact on changes in the bell curve.

Improved Investor Awareness

Improved investor awareness shifts the curve to the right and causes a change in the range of scenarios investors consider. Positive scenarios are weighted with a higher probability of materialization. Moreover, some of the negative development scenarios are no longer considered realistic. Instead, some more optimistic scenarios that entail higher growth and profit levels are now considered more probable (scenario Z). In figure 6-25 this would mean that the bell curve moves to the right.

Increased Investor Confidence

Through comprehensive and real-time provision of relevant company information, the risk perceived by investors can be minimized as well. Market information deficits are eliminated, and shareholder loyalty rises. In addition, the different views various investors hold about the company are harmonized. This homogenization leads to an alignment of investor expectations. In figure 6-25 this results in a thinner and higher bell curve.

Yet another advantage of this type of communication is the prevention of rumors and sensitive information being spread to the public. Speculations can thus be eliminated, which also reduces stock price volatility. This in turn lowers the risk for investors. As a result, they require lower returns from the company. In the *Capital Asset Pricing Model* (box 10) this is demonstrated by the reduction of the company specific risk (ß).

Box 10:   Capital Asset Pricing Model

The volatility of a stock is reflected by the ß-factor in the Capital Asset Pricing Model (CAPM). The cost of equity is the expected minimum return on an investment in a similarly risk-exposed stock portfolio. The ß-factor reflects the company specific risk perceived by the market.

$COE = r_f + ß \bullet (r_m - r_f)$ with

COE: Cost of equity

r_f: risk-free interest rate (e.g., government bonds)

r_m: Expected return of a completely diversified stock portfolio (e.g., total market index)

ß: Factor representing company specific risk (the deviation of the company stock with respect to the diversified market portfolio)

In the DCF-Model (appendix B), too, the company value increases with a reduced risk perception. The reduction of the risk factor by investors reduces a company's capital costs. This means that future cash flows are discounted at a lower average cost of capital. Thus, the current company value is increased.

6.5.2   Investor Communications Audiences and Instruments

Investor awareness is improved if the company systematically uses the available communications instruments and addresses the right audiences.

6.5.2.1 Financial Analysts as Opinion Leaders

Investor communications target a wide range of interest groups with different informational needs. A simultaneous addressing of financial analysts, institutional investors, private investors, and financial publications requires an information policy that is both consistent and also tailored to the specific requirements of these different groups.

Because of their role as opinion leaders financial analysts are one of the primary target groups for successful communications (figure 6-26). This group has a crucial multiplying function, because it issues completely independent investment recommendations for the capital market, which then serve as aids in the decision-making processes of other groups. For instance, the investment decisions of the important institutional investor group is heavily influenced by the recommendations of the analysts, while private investors are frequently following the trends set by institutional investors.

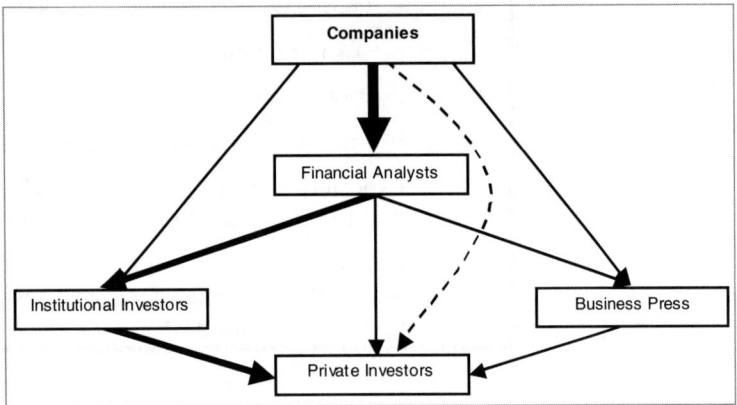

Figure 6-26:   Information flow and resulting interrelations of individual interest groups

Intensive communications with analysts is crucial for another reason, too. The reputation of financial analysts depends in large parts on the accuracy of their predictions about the future quarterly results of companies. The more consistently a company provides analysts with detailed and meaningful information, the better they are able to satisfy their customer's requirements. The company obviously profits from this as well. If a business plan receives credible support from several independent analysts, investor confidence in the communicated company development increases.

Direct contacts with institutional investors can be particularly helpful in influencing price relevant purchases and sales of large stock packages in the run-up.

6.5.2.2 Sustainable Value Creation Through a Credible Equity Story

Investor Com-
munications
Instruments

There are different communications instruments to provide both quantitative and qualitative information. These instruments frequently have a time-dependent focus, i.e., they are either rather long-term or more short-term-oriented. The instruments described in figure 6-27 either have a positive impact on investor awareness or on investor confidence.

	Quantitative	Qualitative

Short-term
(1-5 years)
	• Quarterly reports
	
	• Annual report
	
	• Ad-hoc-
	  publications
		• Annual general
		  meeting of
		  shareholders
		
		• Stock ads
		
		• International IPO
		
		• CEO image

Long-term (Terminal value)	• Analyst meetings • Equity story	• Analyst meeting • Equity story

Figure 6-27:   Crucial investor communications instruments

Short-term instruments are meant to positively display the cash flows of the company in the immediate future. This short-term horizon, which can be predicted with relative accuracy, can range from one year (e.g., growing companies) to five years (e.g., established large corporations). Long-term instruments on the other hand primarily target the increase of the so-called terminal value.

As described earlier, the terminal value accounts for the lion's share of a company's valuation (up to 90 percent in growth industries). Effective investor communications therefore aims primarily at long-range projections.

The Importance
of the Equity
Story

Thanks to its long-term approach, the Equity Story is the instrument that has the most significant impact on terminal value expectations. Therefore it also has the most crucial influence on the total company valuation and serves primarily in communications with the financial analyst target group.

The Equity Story provides a preview of the financial success of a company that it expects in its target markets based on its planned business model. Thus it stipulates the reasons why potential investors should invest into the business: it shows the long-term options and potentials of the company that allow it to continue to grow and to sustainably increase its profitability.

Obviously, the Equity Story must be communicated in a comprehensible manner. Especially in case of companies in emerging industries it is particularly difficult to assess the opportunities and risks of innovative business ideas because there is hardly a basis for comparison. These companies therefore must do whatever they can to convince financial analysts of their potential for success.

Equity Story Confirmation

The Equity Story is rated in comparison to the aspirations of competitors. If a company intends to grow at a faster pace than its current market, it must either gain additional market share or penetrate new markets. Therefore the Equity Story is aligned with the overall market development projections. Analyst meetings have proven effective for the communication, discussion, and validation of Equity Stories.

Investors will only accept the Equity Story as a basis for the valuation of the company if the short-term goals and results derived therein are being met. To verify this, investors review the quarterly results. If these are consistent with the expectations that financial analysts derive from the Equity Story beforehand, the validity of the Equity Story is confirmed.

If the expectations are not met, however, investors begin to have second thoughts about the Equity Story. Long-term result expectations are reduced, and the valuation of the company shrinks. If company results miss market expectations – even only by a few cents per share – this can result in severe share price adjustments.

Quarterly and Annual Report

The quarterly report and ad-hoc publications deliver previously unknown information that has a direct impact on the short-term share price component. While the quarterly report communicates straightforward quantitative data, ad-hoc releases can also contain qualitative information.

Quarterly results are primarily used to confirm the long-term Equity Story in the short term. The annual report, on the other hand, is merely a report of past business developments. Its design and content can, however, be helpful in improving a company's image, and it is also frequently used to distribute specific information.

Other Short-Term Instruments	Another popular investor communications tool is the personification of a company's success through executive portraits. If these successes are, for example, attributed to the abilities of the CEO, the market expects these positive developments to continue in the future. The image of a CEO can indeed enhance the value of a company continuously, as was evident in the case of Jack Welch at General Electric. The appointment of a new CEO can also result in an instant share price leap. In 1998, when Richard Belluzo became chairman and CEO of Silicon Graphics, the stock price increased by more than 20 percent.

Shareholder meetings, stock ads and international IPOs or public offerings in another stock market segment also have trust-building effects. When the German Neuer Markt encountered continuing weakness in 2001, threatening to pull down fundamentally sound companies in its wake, some of those companies considered to move their listings to the more conservative (and lower-risk) mid-cap index MDAX. They intended to remove themselves from the downtrend of this particular market segment. Such a move may also have a direct beneficial impact on the share price, since index funds and index-oriented investors are now increasingly adding this security to their portfolio.

6.5.2.3 Investor Awareness and Accounting Standards

Among other things, investor risk perceptions and return on investment expectations are based on the financial statements. Companies must therefore take into account that different accounting standards have an impact on the risk perceptions of the investors and therefore on the company valuation.

German Accounting Standards	Until just a few years ago, the international stock markets offered German companies only limited funding options. Consequently, many companies funded their investments through loans from a few large banks. These banks were mainly interested in secure and on-time interest and loan repayments. Given these circumstances, creditors used their leverage to influence the design of accounting standards. German accounting standards are therefore designed to protect the interests of creditors.

A high level of creditor protection tends to result in the overvaluation of liabilities and the under-valuation of assets. This cautious approach does, however, distort the actual economic situation and therefore makes it harder for equity capital providers to rate their return on investment options. The diversity of voting rights and accounting leeway adds fuel to the fire.

Some large German companies are changing their German accounting standards to those of the *US GAAP,* primarily to allow their stocks to be traded on the New York stock exchange. While easier access to capital was the prime motivator behind this move, these companies are also being rewarded with increased investor confidence due to the substantially higher level of transparency and more stringent guidelines. US GAAP balance sheets can therefore send a positive signal to the capital markets: it indicates that the company does not have to cover up its economic situation.

6.5.3   Expectation Management is at the Heart of Successful Communications

Every market has certain expectations for publicly traded companies. To enhance the stock market valuation a company must at least meet these market expectations. Applying its communications instruments skillfully, companies can, however, influence these expectations, or even control them.

6.5.3.1 Exceeding Expectations Yields Value Creation

To attain an above-proportional increase in market valuation, a company must surprise the market with positive news that have not yet been factored in the current valuation. As described in chapter 1, above average value can only be created if investor expectations (E^3) are continually exceeded.

In the event of quarterly results that exceed expectations, it is not just the realized higher earnings that are considered in the valuation of the company. The perpetual growth rate is also adjusted to reflect the past, increased performance. This multiplier effect is significant because the share price reflects the capital value of all changes in investor expectations with respect to future free cash flows. The share price takes a leap.

In practice, the capital markets usually anticipate better quarterly results – especially if companies continually inform their investors about earnings projections. Consequently, company valuations frequently already adjust to the projected over-performance between quarterly earnings reports. If a company continuously exceeds expectations its share price increases exponentially. If it meets earnings expectations its share price grows at a constant rate given that investors expect constant returns. If the expectations are not met the capital markets quickly react with corrective price adjustments. If the (lower) expectations are subsequently met the stock price continues to gain at the adjusted "constant" rate (figure 6-28).

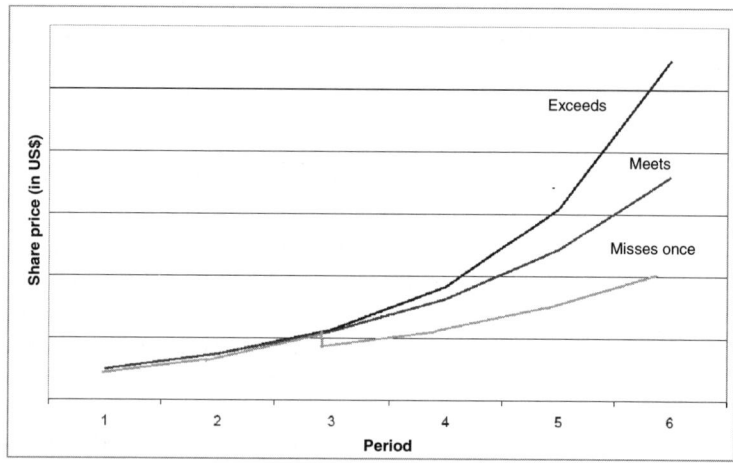

Figure 6-28:   Share price development when expectations are exceeded, met, and missed

6.5.3.2 Managing the Expectation Treadmill

The perpetual growth rate of a company could be seen analogous to an expectation treadmill. A change in the growth rate is consistent with increasing or decreasing the treadmill's momentum. Consequently, unexpected positive results lead to an increase of the expected growth rates and thus accelerate the expectation treadmill. If a company continues to exceed the expectations, it might end up in a vicious cycle: the better its performance, the higher the expectations in terms of future performance. The expectation treadmill continually gains momentum. For the company on the other hand, fulfilling these expectations is an ever more challenging venture.

Companies who are in this situation are therefore caught in a dilemma: on the one hand, they need to create value through results that exceed the market expectations. On the other hand the expectation treadmill must not be sped up too much, so that the company will be in a position to meet future expectations.

Companies that are faced with less demanding expectations find it easier to exceed them in the next period, which allows them to attain above average stock price increases. The danger of disappointing market expectations is also diminished. The non-fulfillment of market expectations is usually subject to severe market corrections. Cisco Systems, for example, had won the admiration of many investors by exceeding expectations by one cent for 14 consecutive quarters. Nonetheless, when Cisco missed its mark by only 1 cent at the end of 2000, its stock price instantly declined by 13 percent.

Companies must therefore work hard to control market expectations by managing their earnings accordingly and by communicating appropriately.

A Harvard Study published by Professor Zeckhauser in 1999, which analyzes more than 100,000 quarterly results in the period of 1974 through 1996, reveals the tactics of successful companies: a quarter of the quarterly results evaluated deviated from the market expectations by just one cent per share. Less than one percent of the results deviated from the analysts' expectations by 20 cents. Results that exceed expectations by one cent per share are far more prevalent than they would be in a random statistical distribution (figure 6-29).

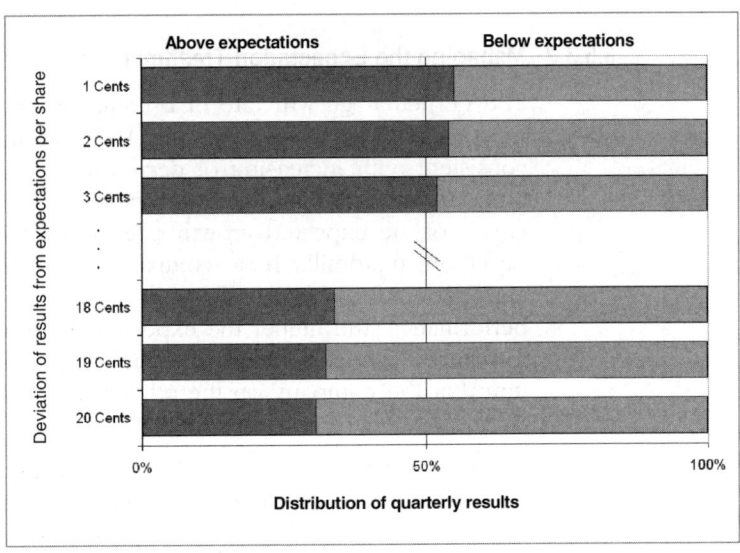

Figure 6-29:   Statistical distribution of deviations of company results from market expectations in cents per share

Always meet or Exceed Expectations Companies obviously strive to meet the expectations of the market, but they aim at not exceeding them by too much. This is due to the fact that the positive relative price lever loses momentum relative to the degree of exceeding expectations. In other words, the maximum effect in both directions is greatest close to the expectations horizon. Consequently, the number of quarterly results that have outperformed expectations substantially is relatively small.

Good communication avoids statements that reveal extreme profit increases in order to prevent future market expectations from becoming too demanding.

If You Have to Miss, Miss by a Mile Negative deviations from expected results are considered a breach of trust by investors and almost always go hand in hand with crashing stock prices. Disappointment is the seed of the uncertainty of whether expectations might be missed even more severely in the future – and even worse – that the equity story may no longer be credible in the long-term.

In this context it is comparatively unimportant if a company misses its market expectations by an inch or by a mile. This is evident in the fact that share prices decline only slightly more severely if the latter is true. In case of doubt, it is therefore better to miss market expectations not just by a few cents, but very noticeably. Companies should exploit this opportunity to bundle bad news and start over with a clean slate. They could, for example get rid of skeletons in the balance sheet to gain new breathing room for future earnings management and to lower the basis of future market expectations. In daily practice, results are therefore frequently revealed that are far below market expectations, which was also evident in the Harvard study cited above.

A lack in corporate communications tends to cause companies to be undervalued by the market. Exaggerated communications can quickly suck a company into a vicious circle, resulting in its failure to deliver on excessive expectations. In consequence, the market punishes this behavior with a reduction in valuation. Intelligent expectation treadmill management via targeted corporate communications can therefore maximize a company's long-term value.

The Complete House:
Holistic Value Creation in Practice

The prior chapters have described the individual elements of the House and presented specific approaches for optimization. It is important to synchronize the individual efforts synergistically . Therefore this chapter will assemble the building blocks in a four-phase process (figure 7-1) to complete the house. Furthermore, the example of GreatValue, Inc. illustrates an integrated value creation program.

The two initial phases of this program summarize once again what was extensively analyzed in the course of introducing the House of Value Creation. Phases three and four focus on the definition and implementation of the value creation program based on the analytical process using the House frameworks. These phases will be laid out in detail in the following chapter.

1. **Identification of Value Gap and Target Setting**

   In order to identify the value gap of a company and to develop a value creation strategy, the capital market valuation needs to be compared to the competition. As already explained in chapter 1, this analysis should take into account the dynamics of the industry as a whole as well as the migration paths of individual competitors. Structural differences between the respective companies also need to be considered to ensure comparability between conglomerates and individual companies. The objective is to determine the capital market value gap and to investigate the underlying causes. Insufficient profitability or slow growth can explain the existence of value gaps at the highest level.

2. **Identification of Barriers to Value Creation and Levers for Improvement**

   Phase 2 determines success criteria for closing the identified gap based on a comparison with competitors. The House of Value Creation is employed to systematically capture and evaluate value creation potentials and to uncover any barriers to their realization. This approach has been extensively described in chapters 2 through 6.

3. **Development of a Comprehensive Value Creation Program**

   A value creation program is now developed from the results of phase 1 and 2. Individual improvement measures are prioritized and integrated into the program according to their relative significance. Aside from individual functional optimization programs such a value creation program consists of efficient program management and control mechanisms with clearly defined time schedules and responsibilities.

4. **Implementation and Monitoring**

   The program management is responsible for the overall coordination of the value creation program. Through the steering committee, the executive level is directly involved in monitoring the progress of the program. Throughout the initiative program goals need to be continuously monitored and adjusted to industry dynamics, if necessary. The objective is the institutionalization of the value creation cycle with feedback to and from the current market situation.

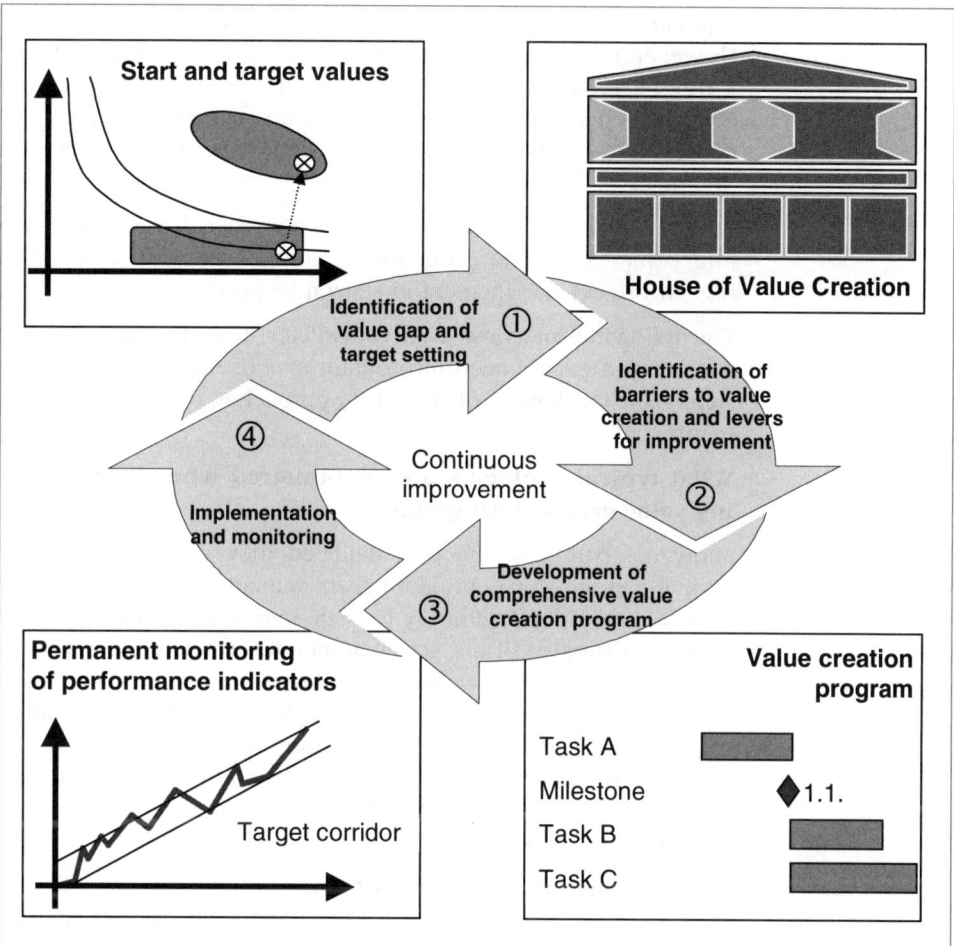

Figure 7-1: Application of the House of Value Creation in the context of a value creation program

7.1 Program Management – A Holistic Implementation Approach

In general value creation potentials can be realized through separate sub-projects confined to the individual business units. However, these projects must be coordinated and integrated into a holistic program in order to ensure the greatest possible success. An effective and efficient program management thus becomes the pivotal point for the realization of all value creation potentials.

Program management represents an approach towards controlling complex change programs in companies that differs from regular project management in several respects.

The following interview with Hans-Peter Remark, Partner at Accenture and experienced in program management, highlights the significant challenges of and the approaches to integrated programs:

What typical problems are encountered when implementing value creation programs?

Many companies in the past believed they could refrain from applying professional program management. In the best case a coordinator without authority to issue instructions would oversee individual projects. Thus, a conglomerate of heterogeneous projects was managed according to different – and often conflicting - metrics and objectives. Lacking coordination caused the implementation to become complex and nontransparent. Overlaps and gaps developed. In spite of the success of individual projects, the whole program, conducted without effective coordination, was often doomed at its inception.

How can such problems be solved?

As soon as several projects are carried out simultaneously and they are aligned to a joint objective, a professional program management becomes indispensable. Program management, in its principle, is really a two-phase project management. The program manager, in other words, acts as the architect or general contractor, who divides the program into several manageable sub-projects, coordinates these sub-projects and manages according to the program schedule. The program management ensures that objectives are clearly defined, frictions avoided, and that all projects contribute to the fulfillment of the program as a whole.

What authority does the program manager require?

The program manager is responsible for the success of the whole program. Therefore he/she has to have extensive coordination and instruction authority. This authority should be applied within the framework of a consistent top-down management to implement and control (monitor) all activities. A program manager defines what quality criteria the achieved results need to meet and monitors the fulfillment of such standards. In addition the responsibilities encompass the planning and assignment of resources as well as comprehensive time and cost monitoring.

What is the usual way to set up a value creation program?

In a first step the program manager defines processes and communicates them to the project managers. Even seemingly obvious aspects should be explicitly described in the program so that all project teams have a common understanding regarding the approach.

In a second step he/she defines the individual sub-projects. The individual functional areas on all program and project management levels have to be clearly defined so that there is no room for interpretations, and so that all relevant aspects of the complete program are processed completely and without overlaps.

Afterwards the resources are allocated to the defined projects and functions. In order to optimize the allocation, the skills required to fulfill the tasks should be described in detail and checked with the capacities available. Therefore it is important to clarify what organizational units can drive which aspects. For the implementation of the program the involvement of line-employees has to be discussed in advance. Time commitments and role assignments have to be detailed. The result of this set-up phase is a program structure that shows what tasks are completed by which employees, with what resources, and within which timeframe.

It is important for reasons of consistency of content and disciplined budget planning to monitor and approve the progress of a complex value creation program in several steps. The steering committee that consists of board members and program management should monitor the results of completed projects after every implementation phase and only then release funds for the subsequent phases.

Do the tasks of program management change over the course of a project?

Value creation programs are often implemented over several quarters or they pass multiple iterations of the four-phase process. The program management therefore must react in a flexible manner to significant changes in the business situation and be capable of adapting program objectives accordingly. For example, this flexibility is severely tested in corporate restructuring situations when a business unit is acquired or disposed of.

Progress monitoring is a significant task of the program management. Because of the long duration of programs, regular project metrics – for instance, the completion level expressed as a percentage – prove to be insufficient. When problems occur the project managers could simply postpone the completion of their tasks. Therefore the program management should only approve completed projects – this procedure is called "earned versus burned" method. But this is only possible when the individual projects are scheduled over a reasonable period of time. According to experience, this period amounts to three to four months. In addition, one advantage of this approach results from the fact that conclusions for the definition of subsequent projects can only be drawn from completed projects.

How can complex programs be successfully implemented company-wide?

If a value creation program leads to changes of organizational units or to the introduction of, e.g., new IT-infrastructure then the effect on the day-to-day business operations can be severe. In this case the implementation needs to be managed very carefully. Test and pilot phases are critically important particularly in case of IT-programs. Accenture plans with several pilot phases for the introduction of large IT-systems and frequently builds in a whole month for trouble shooting and repair. When restructuring the billing systems of Deutsche Telekom this approach helped confirm the error free migration of approximately 40 million customers to the new system.

How can the customer be integrated into the adaptation process?

Communication is one of the significant success factors during the implementation of a value creation program. We recommend to the program managers to build up a pyramid of sponsors within the line-organization. Employees from different levels are involved very closely in the implementation and are always in a position to inform their peers about the objectives, problems and progress of the program. These sponsors are likely to picture the program in a positive light and can therefore positively influence the attitude of their colleagues towards the program. Furthermore, the program management should not neglect direct communication, e.g., via the Intranet or employee television.

Support by the top management is of special importance. Severe adaptation processes such as a value creation program frequently cause costs that amount to millions in large corporations. Typically, only the board or the directly subordinated level can authorize expenditures of this magnitude. The top management must be conscious that a long-term oriented value creation program should not be stopped because of short-term considerations, even if a weak economy demands cost savings. In case of acquisitions, the corporate management should, for the sake of value creation, also see to it that the acquired company's superior processes are also effectively integrated into its own program.

Are there additional advantages of a systematic program management?

A management that oversees all sub-projects of a value creation program has a global overview of a corporation. This perspective often allows for uncovering potentials that would normally remain unnoticed in the complexity of a corporate structure. The extensive overview of the program management can lead for example to the reevaluation of unprofitable product sectors that otherwise would have never been identified as such.

7.2 Case Study – The Example of GreatValue, Inc.

Below, based on the example of GreatValue, Inc., the implementation of a value creation program is presented.

7.2.1 Phases 1 and 2 – Analysis of Value Creation Potentials

As explained in prior chapters, GreatValue determined a valuation gap on the capital market compared to its competitors. As a response, the management declared increasing company value and closing the gap as its foremost objectives.

In the past GreatValue, Inc. had analyzed individual value creation factors with the aim of increasing market capitalization. However, there had not been a comprehensive approach to systematically align all relevant aspects and activities towards the goal of value creation so far. Therefore the management of GreatValue, Inc., introduced the House of Value Creation as the diagnostic instrument.

The aspects shown in figure 7-2 with their respective, specific key questions, were identified as significant value creation potentials. Particularly the lack of growth options outside the core business (bold type in figure 7-2) was explicitly mentioned. Thus, the value creation program focused on this area.

Value creation potentials	Specific key questions
Growth opportunities in core business	• Do we focus our core business on sectors with good opportunities for profitable growth?
Utilization of growth opportunities outside of core business	• **Do we continuously seek opportunities outside of the current core business?** • **Do we consistently evaluate and actively push growth options?** • **Do we support the build up of new business units?**
Renewal of business portfolio	• How do we ensure continuous refreshing of our business portfolio? • Does the market undervalue individual business units due to conglomerate discount?
Operational efficiency in all business sectors and processes	• How well do we manage our customer relationships, especially sales and service? • How quickly and efficiently can we develop products and make them ready for the market? • Is purchasing and production organized efficiently?
Organizational value potentials	• Are we participating successfully in the war for talents? • How entrepreneurial is our organization? • How effective are our investor communication?

Figure 7-2:   Analysis of value creation options at Great-Value, Inc.

GreatValue, Inc., recognized the necessity to create a widespread portfolio of new business ideas. In order to guarantee the innovation capability of the company in the long term, projects with varying time horizons were to be initiated. The monitoring and controls over the projects should constrain the creativity of the employees as little as possible and should therefore primarily focus on milestones. GreatValue created a *value creation office*, a staff position for the implementation of the value creation program. This entity was in charge of the program management as well as coordination of an internal business plan competition.

Search for Opportunities Outside of Core Business

GreatValue, Inc. created a panel of outside experts to expand the informal basis for exchanging ideas about future trends and new business sectors. For the continuous quest for new technological trends, a group of technology scouts was created that reported directly to the value creation office. In addition, all employees were asked to participate actively in company-wide working groups. Their involvement in this area also became a performance evaluation criterion.

Evaluation and Leveraging of Growth Opportunities

The management of GreatValue not only systematically strengthened the access of their company to inter-corporate expertise, but also improved the environment for internal knowledge building. In order to capture and finally to implement valuable ideas, GreatValue, Inc. founded cross-functional working groups in which the employees could exchange trends and ideas on a regular basis. The employees of the value creation office and technology scouts were also invited to participate in these working groups. During the regular meetings ideas were generated that were subsequently supported by the business plan competition process. Aside from that, the technology scouts evaluated interesting start-ups and potential acquisition candidates. These evaluations also turned up prospective knowledge champions to be recruited by GreatValue, Inc.

In order to evaluate both internal and external ideas, an internal business plan competition was set up where content and also possibly the financial implications or opportunities of these ideas could be presented. Sufficient funds were provided in the budget for this purpose. Employees were motivated to participate because of the opportunity of playing a decisive role in a new project and the related career opportunities. This procedure not only allowed the potential assessment of individual ideas, but also the comparison between developing internal ideas versus external acquisitions.

Leveraging the Build up of New Business Units

Ideas considered feasible during the business plan competition were to be subsequently developed by project teams. A model was created that systematically propels projects towards the commercial stage. In this context future customers should be involved in the development phase and an extensive market pilot program carried out. Funding should not happen all at once, but instead was tied to the completion of particular milestones.

It was especially important for GreatValue to build up a wide spectrum of new business ideas. Particular focus was placed on initiating projects with varying time horizons – to ensure the long-term innovation prowess of the corporation. The monitoring and oversight over the projects by the management should be largely limited to the milestones, so that the creativity of the teams would not suffer.

Impact of these Measures on GreatValue, Inc.

Initial improvements at GreatValue, Inc. already became apparent before the completion of the value creation program. New-born entrepreneurship began to mold the corporate culture of GreatValue. Employees collaborated in seeking future opportunities. Numerous highly promising projects were initiated within the shortest period of time.

After these very positive results in its attempt to increase company value by developing new business areas, GreatValue, Inc. is currently identifying additional sectors for future value creation programs. The management is convinced that it possesses the appropriate approach in the form of the House of Value Creation to uncover additional value gaps in the future and to consistently align all business areas towards value creation. Initial successes, as well as the decision to further focus the company on value creation, were communicated directly and openly to investors and analysts. The stock market commended this transparency. The valuation gap versus the competition was considerably reduced within only six months.

8   Outlook – Strengthening the Corporate Future

Economic
Development

Since the second quarter of 2000 the economic environment has been anything but bullish. Massive losses, particularly in the tele-communications, media and technology industry resulted in major slumps of stock market indices worldwide. From peaks in March 2000 the Dow Jones Industrial Average had lost more than 25 percent by October 2001, the German Stock Index had dropped by almost 50 percent during the same period and the technology stock exchange NASDAQ experienced a downturn of nearly 70 percent. Meanwhile, the Germans growth stock index, Nemax All Shares, saw the biggest slump at more than 90 percent.

The weak global economy played a crucial role in these developments. After the United States had enjoyed the longest sustained economic boom in history, it is now in the throes of a recession. Japan saw substantial losses as well and Europe is watching its dynamic growth getting sluggish.

As was evident in stock market indicators, high tech stocks were the biggest losers in the last year and a half. While so-called "Old Economy" stocks were also affected by this trend, they were able to stave off a downtrend and protect their market values until the second half of 2001.

The current economic climate is causing many companies severe headaches. Numerous high tech companies failed to meet analyst expectations for the very first time and suffered punitive consequences: their stock prices fell. News of profit warnings and declining revenues dominated company press conferences in 2001. Consequently, massive layoffs began, investors withdrew capital and further industry consolidations were initiated.

The outlook for 2002 is therefore rather mixed. Leading experts do not expect a turnaround until the second half of that year. The hope remains that the U.S. economic engine will jump-start and that some of the European countries that have shown strong growth potential will be able to expand on their strengths.

What can companies do in such times of an economic downturn?

Options for Companies

The House of Value Creation shows that companies can stand strongly against negative capital market trends and increase their market valuations.

Therefore, business leaders should primarily focus on identifying valuation gaps, tracking down their causes and selecting an ideal combination of value creation levers. These levers must first and foremost strengthen the foundation of the company and align employees, business units, business architecture, portfolio structure, and corporate finance with value creating objectives. The House of Value Creation provides a tool that supports managers at all levels in achieving these goals.

The value creation levers growth and profitability can help enterprises, regardless of their size and industry, to improve their stock performance under all economic conditions. This in turn helps create investor satisfaction and thus incentives for new investors. Companies who utilize these tools are better equipped to defend themselves against hostile takeovers while increasing credit ratings and options to borrow capital at the same time. Consequently, innovative ideas can be commercialized more quickly and more effectively. In the long-term, all of this benefits the survival and independence of the company. Hence, the cycle embarked upon at the beginning of this book is now complete: shareholder value creates stakeholder value.

The authors hope that you, the reader, will find our overview and evaluation of the various approaches and methods helpful and are very much looking forward to receiving your feedback at www.house-of-value-creation.com.

Appendix

A: Value Creation – Relative Valuation Map

The relative valuation map compares the market capitalization with sales. The sales multiple thus produced provides the relative market valuation of a company. In our computations sales were used instead of book values, since sales are the more appropriate parameter to apply to dynamic industries with numerous mergers & acquisitions and different starting points in terms of companies' equity.

In the relative valuation map (figure A) lines of identical market capitalization, i.e. various combinations of sales and relative market valuation are shown as value isoquants. The positioning of a company on this map allows an assessment of its current competitive situation as well as its strategic reach (for example, active market impact potential through acquisitions). Generally, companies can be classified into four groups:

- **Value Leaders** are companies characterized both by high sales multiples and high sales. They can influence industry dynamics and competitive structures due to size as well as value.

- **Climbers** are companies that have achieved high valuations while their sales are (still) small. The high sales multiple reflects high expectations. Most of the companies in this segment are young, highly innovative enterprises.

- **Incumbents** are generally traditional businesses with solid sales and relatively low market capitalization.

- **Center Players** encompass companies with low sales and low market capitalization.

This categorization is applicable across industries, while not all groups may necessarily be clearly distinguishable in all industries.

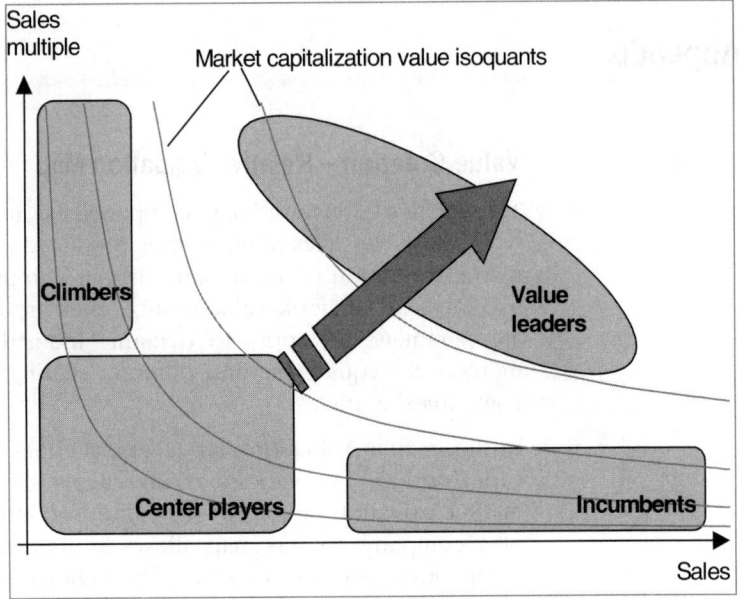

Figure A: Relative valuation map

Different migration paths lead to higher market capitalization. Hence, most companies show heterogeneous strategies in terms of prioritization of growth and profitability. Their movements in the relative valuation map over time do, however, permit an evaluation of their value creation performance regardless of the company focus.

B: Using the DCF Method to Derive Company Values

One method of computing a company's value is the application of the Discounted Cash Flow (DCF) method. According to this method, the value of a company (V) is the sum of the discounted, free cash flows within an explicitly defined forecast period plus a residual value called Terminal Value (TV) that is also discounted to the present. The latter consists of a perpetuity with constant growth rate. Expressed as an equation this reads:

Equation 1   V = DCF + TV

Mathematically this equation translates into:

Equation 2

$$V_{t=0} = \sum_{t=0}^{T} \frac{FCF_t}{(1+WACC)^t} + \frac{NOPLAT_{T+1}\left(1 - \dfrac{g}{ROIC}\right)}{(WACC - g) \times (1 + WACC)^{T+1}}$$

Definitions:

V:	Company value
FCF:	Free cash flow
WACC:	Weighted average cost of capital
NOPLAT:	Net operating profit less adjusted taxes
g:	Perpetual NOPLAT growth rate
T:	Final year of the explicit forecast period
ROIC:	Return on invested capital
COE:	Cost of equity

For corporations listed on the stock market, an external company valuation exists in the form of its market capitalization (M), which is equal to the expression for V:

Equation 3

$$M_{t=0} = \sum_{t=0}^{T} \frac{FCF_t}{(1+WACC)^t} + \frac{NOPLAT_{T+1}\left(1 - \dfrac{g}{ROIC}\right)}{(WACC - g) \times (1 + WACC)^{T+1}}$$

This Equation implicitly reflects investor expectations in terms of growth (g) and profitability (NOPLAT), given that the freely available cash flow in (2) and (3) is calculated based on the pre-tax operating profits (EBIT – Earnings Before Interest and Taxes) as follows:

	EBIT
-	Taxes
=	NOPLAT

	NOPLAT
+	Depreciations
-	Increase in current assets
-	Investment expenses
=	FCF

Consequently, FCF could also be interpreted as the non-invested portion of NOPLAT:

Equation 4 $$FCF = NOPLAT \times (1 - IR)$$

IR denotes the investment rate. The expected growth rate (g) is the product of investment rate and expected return on investments:

Equation 5 $$g = ROIC \times IR \Rightarrow IR = \frac{g}{ROIC}$$

Equations 4 and 5 combine to:

Equation 6 $$FCF = NOPLAT \times \left(1 - \frac{g}{ROIC}\right)$$

If equation (6) is entered into equation (3) the result is:

Equation 7

$$M_{x=0} = \sum_{t=0}^{T} \frac{FCF_t}{(1 + WACC)^t} + \frac{FCF_{T+1}}{(WACC - g) \times (1 + WACC)^{T+1}}$$

The company valuation therefore consists of the DCF component, which describes the near future and the Terminal Value component. Parameter T defines the end of the explicit forecast period and thus separates the two components. The accuracy of the calculation hinges on how realistically a prognosis for the FCF values in the forecast period can be made. Usually, the business plan of the company evaluated is consulted first, since it generally contains the relevant figures for the upcoming years. This time period can then be extended (depending on prognoses about market and company development) by a number of years.

The period for which freely available cash flows can be predicted relatively reliably very much depends on the industry in which the evaluated company operates. In very dynamic industries, this period usually does not exceed two to four years. Of course, the relative distribution of the company's value to the DCF component and the Terminal Value Component also depends on the length of the forecast period. Overall, the Terminal Value prevails in all industries. It is, however, not uncommon for this residual value, which is largely dependent on investor expectations, to be responsible for more than 90 percent of the market capitalization, especially when this kind of analysis is applied to emerging "New Economy" companies. In many cases where FCF values are predominantly negative during the forecast period, the Terminal Value may actually exceed the total value of the company. One typical example of such a New Economy corporation is Yahoo, one of the world's leading Internet portal operators. In June 2000, the discounted cash flows during the forecast period accounted for a mere 3 percent of the overall company valuation, 97 percent was based on capital market expectations (figure B).

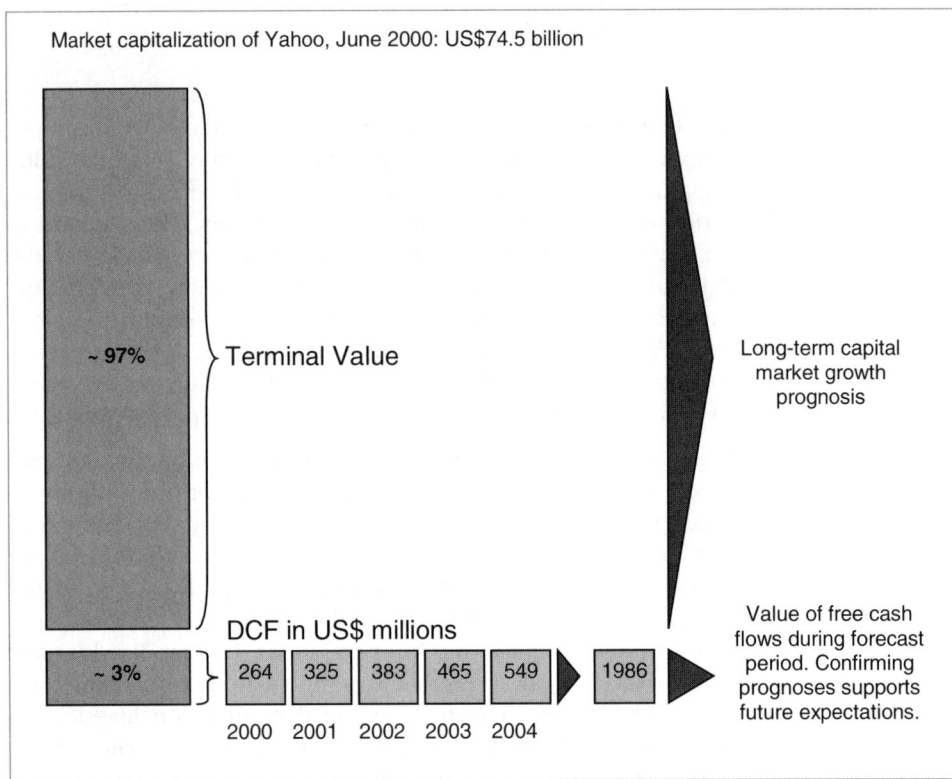

Market capitalization of Yahoo, June 2000: US$74.5 billion

~ 97% } Terminal Value

~ 3% }

DCF in US$ millions

| 264 | 325 | 383 | 465 | 549 | | 1986 |

2000 2001 2002 2003 2004

Long-term capital market growth prognosis

Value of free cash flows during forecast period. Confirming prognoses supports future expectations.

Figure B: Example of company value distribution to DCF and Terminal Value components: Yahoo, Inc. (June 2000, source: analyst reports)

In evaluating dynamic industries, where company valuations are largely contingent upon expectations, equation (7) can be simplified by actually neglecting the DCF component. This assumption reduces equation (7) to:

Equation 8 $$M_{t=0} \approx \frac{FCF_{T+1}}{(WACC - g) \times (1 + WACC)^{T+1}}$$

If one were to view the company merely from an equity stand-point, yet another simplification could be applied, since the weighted average capital cost can now be replaced by the COE:

Equation 9
$$M_{t=0} \approx \frac{\text{FCF}_{T+1}}{(\text{COE} - g) \times (1 + \text{COE})^{T+1}}$$

The free cash flows can be defined as the product of revenues (R) and a profitability index (π), which further reduces (9) to:

Equation 10
$$M_{t=0} \approx \frac{R_{T+1} \times \pi_{T+1}}{(COE - g) \times (1 + COE)^{T+1}}$$

If (10) is divided by R_{T+1}, the sales multiple SM appears on the left-hand side, which is derived by dividing market capitalization by revenues:

Equation 11
$$SM = \frac{M_{x=0}}{R_{T+1}}$$

Thus:

Equation 12
$$SM \approx \frac{\pi_{T+1}}{(COE - g) \times (1 + COE)^{T+1}}$$

In other words, the relative valuations of companies in dynamic industries is predominantly explained by market expectations in terms of the two parameters perpetual profitability (π) and growth rates (g) under the assumption of similar costs of equity.

Table of Figures

Figure 1-1:   Dimensions of the capital market analysis..........6

Figure 1-2:   Development of the Nemax All Shares
                and Nemax 50 indices 8

Figure 1-3:   Number of winners and losers by industry.........9

Figure 1-4:   Derivation of E^3 from the Total Return to
                Shareholders..11

Figure 1-5:   Performance indicators of the four high-
                tech industry segments analyzed......................12

Figure 1-6:   Relative valuation map of the Network
                Suppliers and Industrial Electronics
                segments in June 2001 13

Figure 1-7:   Relative valuation map of the
                Semiconductors and Computers Segments
                in June 2001 14

Figure 2-1:   Blueprint of the House of Value Creation 20

Figure 3-1:   Correlation between the visions and
                achievement at NEC and Sun
                Microsystems 28

Figure 3-2:   Vision and value creation at Microsoft.............29

Figure 3-3:   Individual aspirations vs. business plan at
                GreatValue, Inc. 35

Figure 3-4:   GreatValue: Evaluating the barriers..................37

Figure 3-5:   Business plan with 2.3 percent sales
                growth and 13.0 percent profitability...............39

Figure 3-6:   Comparison of business plan and
                aspirations of senior management....................40

Figure 3-7:   Market value of GreatValue compared to
                competitors..41

Figure 3-8:   Goals of the GreatValue workshop to
                align aspirations 43

Figure 4-1:   Dependence of the impact of the value
                creation levers on the capital market
                valuation...48

Figure 4-2:   Schematic graph of the benchmarking
                between GreatValue's business units vs.
                those of main competitors 52

Figure 4-3:   First phase of a value creation program 54

Figure 4-4:   Second phase of a value creation program 55

Figure 4-5:   Phases of the profitability improvement
                program 56

Figure 4-6:   Steps for detailing a profitability gap 57

Figure 4-7:   Top-down comparison of cost structures
                per unit produced 58

Figure 4-8:   Comparison of own cost structure to a
                competitor using the benchmarking matrix 60

Figure 4-9:   Comparison of wage levels and
                productivity of selected countries 62

Figure 4-10:   Optimization potentials resulting from
                functional benchmarking 64

Figure 4-11:   Starting points for a reduction of material
                costs 65

Figure 4-12:   Exemplary estimate of cost reduction
                potential of material costs 67

Figure 4-13:   Calculation of the optimization potential 70

Figure 4-14:   Overview of a profitability improvement
                program 73

Figure 4-15:   Assignment of completion levels to the
                phases of the profitability improvement
                program 74

Figure 4-16:   Elements of the program monitoring 76

Figure 4-17:   Structuring the growth pipeline over
                horizons 78

Figure 4-18:   Requirements for the business pipeline to
                ensure an adequate business portfolio in
                the long-term 81

Figure 4-19:   Different requirements of the three growth
                horizons 82

Figure 4-20:   Investment activity at GreatValue, Inc.
                compared to main competitors 86

Figure 4-21:	Program for the identification of potential new business areas	91
Figure 4-22:	Parameters for the identification of potential new business areas for Great-Value, Inc.	92
Figure 4-23:	Filtering process to reduce the entire search space to an initial selection	93
Figure 5-1:	Transfer price matrix	102
Figure 5-2:	Measuring and Controlling Methods	104
Figure 5-3:	Typical metrics	106
Figure 6-1:	Correlation of organization, processes and systems within the business architecture	116
Figure 6-2:	Selection of optimization measures	118
Figure 6-3:	Process optimization phases	119
Figure 6-4:	Idea evaluation and selection	120
Figure 6-5:	Creating shared services	122
Figure 6-6:	Phases of an eReporting system implementation	124
Figure 6-7:	Project selection using a filtering procedure	127
Figure 6-8:	Estimation process methodology	129
Figure 6-9:	Earned versus Burned method	131
Figure 6-10:	Interface problems due to lack of product management	132
Figure 6-11:	Product management involvement and shared responsibility	133
Figure 6-12:	Material cost savings potentials	134
Figure 6-13:	Phase model of an eProcurement implementation	135
Figure 6-14:	Elements of product design optimization	137
Figure 6-15:	Product design optimization phases	137
Figure 6-16:	Outsourcing matrix facilitates decision making	139
Figure 6-17:	Phases of efficient supply chain management	141
Figure 6-18:	CRM processes and target groups	143

Figure 6-19:   Profitability improvement potential
                 through CRM systems 145

Figure 6-20:   Capital costs in correlation to national
                 capital market liquidity 147

Figure 6-21:   Advantages and disadvantages of
                 purchasing options... 154

Figure 6-22:   Stages of an employment relationship 157

Figure 6-23:   Scheme of a balanced scorecard 161

Figure 6-24:   Aspects of implementing a central
                 knowledge management system 165

Figure 6-25:   Probabilities of company development
                 scenarios.. 170

Figure 6-26:   Information flow and resulting
                 interrelations of individual interest groups 172

Figure 6-27:   Crucial investor communications
                 instruments 173

Figure 6-28:   Share price development when
                 expectations are exceeded, met, and
                 missed.. 177

Figure 6-29:   Statistical distribution of deviations of
                 company results from market expectations
                 in cents per share.. 178

Figure 7-1:   Application of the House of Value
                 Creation in the context of a value creation
                 program 183

Figure 7-2:   Analysis of value creation options at
                 GreatValue, Inc. 189

Figure A: Relative valuation map................................... 196

Figure B: Example of company value distribution to
                 DCF and Terminal Value components:
                 Yahoo, Inc. (June 2000, source: analyst
                 reports) 200

Box 1:   Exceeding Economic Expectations............................. 10

Box 2:   And What About the Strategy?................................... 30

Box 3:   Productivities in Different Countries 61

Box 4:   The Business Plan... 89

Box 5:   Myths About Post-Merger Integration 95

Box 6:   The Industry Clockspeed 98

Box 7:   Strategic Planning & Portfolio Management 109

Box 8:   Tenovis Case Study 125

Box 9:   How Effective is the Reduction of Capital Em-
           ployed? 150

Box 10:   Capital Asset Pricing Model 171

Glossary

Aspirations	Aspirations are goals of the individual – on personal, corporate and/or departmental levels as well as in terms of personal contributions and role definitions.
ß-Factor	The ß-Factor is the company-specific risk perceived by the market within the Capital Asset Pricing Model (CAPM). It indicates the relationship between the stock price development and that of an overall market index and is consequently utilized as an index for the stock's sensitivity to capital market movements.
Balanced Scorecard	The Balanced Scorecard is a corporate and employee management tool. In this "weighted objective scorecard" quantitative and qualitative measures are recorded and rated.
Benchmarking	Benchmarking refers to the comparison of indicators. In this case, a company's own indicators (e.g., costs per employee) are compared to suitable external figures. The leading comparative indictor is regarded as the benchmark to which the company compares its own results. This method is used to determine the maximum potential for improvement.
Business Architecture	The business architecture encompasses a company's organization, processes, and systems.
CAGR	Compound Annual Growth Rate (CAGR) refers to the cumulated annual growth rate.
CAPM	Capital Asset Pricing Model (CAPM) is a formula used to determine cost of equity while taking the risk premiums on securities into account. The volatility of a stock is reflected in the ß-factor. The expected rate of return is the expected minimum return on an investment in a stock portfolio with similar risk. $ERR = r_f + ß \bullet (r_m - r_f)$ with ERR: Expected rate of return, serves as a proxy for Cost of Equity (CoE) r_f: Risk-free interest rate (e.g., government bonds),

r_m: Return on investment of a completely diversified stock portfolio (e.g., overall market index),

ß: Statistic indicator for the deviation of a company-specific risk from an investment into a diversified stock portfolio.

COE	Cost of Equity (COE) refers to the minimum returns on investment expected by investors. See Capital Asset Pricing Model (CAPM).
Conglomerate Discount	A conglomerate discount on the market valuation is the result of insufficient conglomerate transparency for the capital markets and related potential cross-subsidization provided to loss-bearing operations.
Corporate Vision	A corporate vision provides either a concrete or an abstract description of a company's goals, core business and strategic focus.
CRM	Customer Relationship Management (CRM) refers to a consistent corporate approach that is driven by the needs of the customer, supported e.g. through Internet presence or customer interaction centers (call centers).
DCF	Discounted Cash Flow – According to the DCF method, the value of a company is the total of the discounted free cash flows (DCF) within an explicitly defined forecast period and the discounted residual value called Terminal Value (TV).
Earned Versus Burned	Earned versus Burned describes a method to continuously verify the progress of projects, especially in the area of research and product development. Costs incurred at the time a milestone is reached are compared solely to the results achieved at that time. This prevents companies from underestimating the required remaining investments in terms of time and money.
EBIT	Earnings Before Interest and Taxes – This term refers to a company's pre-tax, pre-interest earnings and does not take extraordinary earnings into account. By eliminating these three factors, a comparable value is obtained, indicating the actual operational earnings potential of an enterprise, regardless of the company's specific capital structure.

EBITDA	Earnings Before Interest, Taxes, Depreciation, and Amortization (EBITDA) is the same as the EBIT minus depreciations and goodwill amortizations. EBITDA is widely used internationally and is one of the best performance indicators for the operational earnings potential of businesses. Given that international companies adhere to different accounting standards, e.g. due to different statutory requirements, the EBITDA index allows more meaningful comparisons in terms of earnings potential than indicators based on net income.
Economic Profit Plus	Economic Profit Plus is comparable to EVA (see below). It does, however also take long-term effects into account (e.g., savings systems or innovation rates).
Equity Story	The equity story is a preview of a company's expected financial performance in its target markets based on the planned business model and market dynamics.
ERP System	An Enterprise Resource Planning System is an administrative business software system. It allows companies to automate and integrate a large portion of their business processes, while utilizing common data and processes throughout the organization. It also supports generation of and access to information in near real time.
Eurobond Markets	Eurobond markets provide an alternative funding option. These markets are typically not restricted by government policies, since loans are not issued in local currencies. There are no reserve requirements and thus banks do not have to factor them into the interest rates as opportunity costs. Consequently, this system allows companies to obtain cheaper loans and reduce their third party capital costs. Nonetheless, obtaining loans from Eurobond markets is also more risky, given that the minimum reserves in other markets do guarantee bank liquidity. As is the case with all international funding methods, the currency fluctuation risk must be taken into account as well and it may be prudent to install respective safeguards.
EVA	Economic Value Added (EVA) is a control instrument used to evaluate a company's entrepreneurial performance. EVA = net profits − capital costs

Exceeding Economic Expectations (E^3)	E^3 measures the extent at which a company exceeds investor expectations in stock performance:
	E^3 = Total return on investment – cost of equity capital – residual market risk
Factoring	Factoring refers to the sale of a company's receivables prior to their due date. These funds are sold to a factoring company in an effort to improve corporate liquidity.
Fronting Loans	Fronting Loans are a form of capital transfer. In this case a group company grants a loan to another group company. A bank that operates offices in both countries (if applicable) acts only as a broker to complete the transaction. From a legal standpoint, the companies enter into two separate contracts with the bank. One of the companies invests a certain amount with the bank and the bank releases the same amount, deducting a small fee, to the company in the other country. This construction allows companies to bypass capital flow restrictions while taking advantage of tax and interest rate advantages.
Goodwill	Goodwill refers to immaterial assets and is usually used synonymously with company value. In case of acquisitions it describes the difference between the purchase price and the book value of the company being acquired.
Growth, External	External growth refers to growth through mergers, acquisitions and alliances.
Growth, Organic	Organic growth refers to internal growth, i.e., through geographic expansions, new product developments and new distribution channels.
Hybrid Analysis	A hybrid analysis establishes the relevant cost leader for each cost pool or corporate function. The fictional company that is thus created consists only of the best of each function and is referred to as a hybrid.
IPO	An Initial Public Offering (IPO) usually goes hand in hand with the capital market approval of stock capital and the beginning of a company's listing on the stock market. From a company's point of view, an IPO refers to the acquisition of external growth capital using stocks as a funding instrument.

Incentive System	Incentive systems are monetary and non-monetary compensations handed to employees based on the achievement of specific results or the compliance with desired behavioral patterns.
Industry Clockspeed	According to the industry clockspeed concept, industries are driven at different evolutionary speeds. One differentiates between process technological, product technological and non-technological (e.g., organizational) components.
LBO	Leveraged Buyout (LBO) is the takeover of a majority stake in a company. Third party capital is employed to make the purchase.
Leverage Effect	The Leverage Effect refers to the additional impact (leverage) of profitable investments. Combined with a favorable interest on loans this results in a positive impact on the return on equity. If the return on an investment is higher than the interest paid for borrowed capital, the investor has the opportunity to achieve higher profits by borrowing additional capital, which in turn increases the return on equity.
Market Capitalization	Market capitalization reflects the current stock market value of a company. It can be calculated by multiplying the current stock price with the total number of (outstanding) shares.
MBO	Management Buyout (MBO) refers to the takeover of a company by its existing management.
NOPLAT	Net Operating Profit Less Adjusted Taxes – profit after taxes (includes tax adjustments).
Pooling of Interest	Pooling of Interest is an accounting principle, which can be applied during mergers and acquisitions. In this case, the balance sheet items of both companies are simply added up without taking the market value, i.e., the goodwill of the company to be acquired into account. Therefore, the accounting results of the company are not being impacted by goodwill amortizations.
Price Cost Gap	In order to define the target costs in the context of a profitability improvement program the dynamics of pertinent parameters have to be taken into account. Potential factor cost increases and price declines that may occur in the period up until the target year are called price - cost gap. They usually increase the need for corrective action in terms of efficiency improvement.

Profit Center	Profit centers are corporate units that are being managed on a market- and value- oriented basis and evaluated/rewarded accordingly. They are basically companies within companies and have clearly defined objectives, tasks, rights, obligations and a wide spectrum of autonomy.
Relative Valuation Map	The relative valuation map plots the revenue multiple vs. the revenue of a company at a given point in time. The position on this map determines the relative capital market valuation of a company. Companies can belong to one of four groups: Leaders, climbers, center players and incumbents. This categorization can be applied to all industries.
Revenue Multiple	A revenue, or sales multiple refers to the multiple of the current revenue the capital markets are willing to pay for the stock capital of a company (market capitalization/revenue). This is particularly helpful for the valuation of companies if due to a loss situation a price/earnings ratio cannot be established.
RMR	The Residual Market Risk (RMR) includes for example interest and currency exchange fluctuations, speculative stock market rallies or crashes, which are beyond the control of the management.
ROS	Return On Sales = return on revenues, often expressed as a percentage.
Sale-and-Lease-Back	The Sale-and-Lease-back method is a special leasing system. In this case, companies sell assets to a leasing company only to lease it back from that company for continued use. The idea is to attain additional liquidity.
SCM	Supply Chain Management (SCM) refers to the efficient management of the entire logistics chain.
Shared Services	Shared Services are competitive, service and market-oriented organizational units, which are independently accountable for their own results. They usually offer centralized services to internal clients. Typical services rendered are financial, IT, purchasing, and human resource services.
Spin-off	A spin-off refers to the separation of a business unit from a company, eventually followed by a sale or IPO.

Strategic Planning	Strategic planning defines the process of adapting a company's objectives and resources to changing market opportunities.
Tactical Gap	A tactical gap is the gap that usually exists between corporate strategies and implementation initiatives and programs. From it, businesses can derive the key elements required to make a strategy operational.
Terminal Value	According to the DCF method, the value of a company is the sum of the discounted, freely available cash flows (DCF) within an explicitly defined forecast period and a residual value called Terminal Value (TV), discounted to the start date. The residual value is the present value of a perpetuity with a constant growth rate.
Total value of shares traded	The total value of shares traded is the total amount of all closed transactions pertaining to a security, or the overall market, generated within a certain period of time, or by a specific cutoff date.
TPM	Total Productive Maintenance (TPM) is a management concept that is effective on a company-wide basis. By systematically and continually improving the maintenance of production equipment, it maximizes productivity, quality, economy of scale and workplace safety of a company.
TQM	The Total Quality Management (TQM) concept places quality at the center of a company's focus. It may be customer, process or value creation driven. All employees are accountable for the quality attained at their specific level.
Tracking Stocks	Tracking Stocks are the securities of an individual business unit, issued in the name of the parent company. The return on investment of these securities reflects only the performance of that unit.
Transfer Price	Transfer prices refer to a system of intra-company pricing mechanisms for goods and services. Corporate units invoice prices for goods they ship to each other or services they perform for each other. These prices may be calculated based on cost plus, variable costs, marginal costs or market price and are either negotiated between the units or stipulated by a central pricing function.

TRS	Total Return on Shareholders (TRS) refers to the percentage change in stock prices less disbursed capital flows to investors (e.g., dividends, free stocks).
Unique Selling Proposition	A competitive advantage considered valuable by the customer is a unique selling proposition.
US GAAP	United States Generally Accepted Accounting Principles – US GAAP are the American accounting standards issued by the independent Financial Accounting Standards Board, FASB since 1973. They are comparable to the International Accounting Standards but very different from the German HGB (German Trade Statutes).
Venturing	Venturing is the provision of funding to high risk ventures, i.e., promising start-ups (refers to external as well as internal funding).
WACC	Weighted Average Cost of Capital (WACC) is the average capital cost rate for equity and borrowed capital.

Index

A

accounting standards · 175
acquisitions · 152
aspirations · 20, 27, 31
   harmonization · 32

B

balanced scorecard · 23, 107, 160
benchmarking · 19, 57
   functional · 63
   matrix · 59
beta-factor · 171
business architecture · 24, 94, 115
business plan · 35, 50, 89
business plan competition · 190

C

capital
   costs · 9
   external · 151
   internal · 149
   management · 148
   sources · 147
   structure · 146
Capital Asset Pricing Model (CAPM) · 10, 171
career management · 162
cash offer · 155
check ideas · 120
communities · 86
company
   culture · 164
   value · 48
completion level · 74
compound annual growth (CAGR) · 40
conglomerate · 28
   discount · 99, 111
corporate finance · 24, 146
corporate goals · 29, 97
cost
   drivers · 59
   optimization · 55
   pools · 59, 70, 116
   reduction initiatives · 70
   semi-fixed · 71
cost of capital · 50, 146, 151
cost of equity (COE) · 10, 197
Cultural School · 30
customer relationship management (CRM) · 24, 143

D

data mining · 144
debt-equity ratio · 146
decision makers · 33
design-to-cost · 21, 66, 136
development scenarios · 169
discounted cash flow (DCF) · 41, 47, 171, 197
drill-down analysis · 57, 69
drop ideas · 120
dynamic adjustment of prices and costs · 54

E

earned versus burned · 130

Earnings Before Interest and
Taxes (EBIT) · 198
EBITDA · 152
Economic Profit Plus · 105
Economic Value Added (EVA) ·
1, 10, 104, 150
economies of scale · 71, 121
efficiency improvement · 115
employee retention · 163
ePlanning / eReporting · 123
eProcurement · 134
equity story · 24, 173
ERP system · 124
estimation process · 128
Eurobond markets · 148
Exceeding Economic
Expectations (E³) · 10
expectation management · 176
expectation scenarios · 169
expectation treadmill · 177
experience-based assumptions ·
68

F

factoring · 150
filtering · 92, 128
financial analysts · 171
fixed price offer · 155
fronting loans · 148

G

go ideas · 120
goodwill · 152
growth · 21, 77
    external · 84, 90, 152
    funding · 149
    horizons · 77
    option pipeline · 77, 80
    options · 82
    organic · 82, 86

portfolio · 80
strategy · 16
venturing · 84, 88

H

high-tech industry
    analysis · 6
human capital · 24, 156
hybrid analysis · 63

I

incentive system · 22, 97, 105
industry affiliation · 5
Industry Clockspeed · 98
information campaigns · 159
internships · 159
interviews · 33
inventory range · 142
investor awareness · 170, 175
investor communication · 24,
169
investor confidence · 170
investor expectations · 10, 169,
176
IPO · 111
IT infrastructure · 115

K

knowledge management · 165

L

Leveraged Buyout (LBO) · 153
leverage-effect · 146

M

management
  individual · 22
  institutional · 22
  portfolio · 109
Management Buyout (MBO) ·
  153
market capitalization · 7, 12
market valuation · 109
mergers-of-equals · 96
metrics · 16, 22, 97
migration path · 14
monitoring · 74
  budget · 75
  measure · 75
  program · 75
  program monitoring office ·
  75

N

Neuer Markt · 5, 7
  analysis · 6
NOPLAT · 197

O

operationalization · 97
optimization
  potential · 57, 70
optimization approaches · 116
organization · 24, 115
organizational foundation · 23,
  109
outsourcing matrix · 139

P

payment terms · 150
performance deficits · 167

performance indicators · 57, 98,
  103
personnel management · 156
pooling of interest · 152
portfolio
  market attractiveness-
    competitive advantage ·
    110
  market share-market growth
    · 110
  optimization · 112
  project · 126
  structure · 23, 109
post-merger integration · 94
price reduction · 65
price-cost gap · 54
processes · 24, 115
  optimization · 119
product design optimization ·
  136
product management · 132
productivity · 61
product-market combinations ·
  110
profit centers · 100
profit gap · 53
profitability · 15, 21
  increasing · 55
profitability improvement
  program · 53
program management · 182
program monitoring · 73
progress monitoring · 186
project management · 184
pyramid of sponsors · 187

R

recruitment processes · 159
relative valuation map · 13, 47,
  195
residual market risk (RMR) · 10
resource planning · 128

return on sales (ROS) · 35
rule 80/20 · 69

S

sale-and-lease-back · 150
sales multiple · 42, 47
sensitivity analysis · 51
Shared Services · 121
shareholder value · 1
spin-off · 111
stakeholder value · 1
steering committee · 73, 185
stock trade · 153
strategic alliances · 84
strategy · 27, 30, 109
sum of the parts model · 13
supply chain management
    (SCM) · 24, 141
synergy · 152
systems · 24, 115

T

tactical gap · 1, 30
target costs · 70, 71
technology scouts · 190
terminal value · 34, 41, 48, 173
termination procedures · 166
Total Productive Maintenance
    (TPM) · 119
Total Quality Management · 19
total return to shareholders
    (TRS) · 10
tracking stock · 112
trading volume · 7
transfer price matrix · 102
transfer prices · 100

U

unique selling proposition
    (USP) · 86
US GAAP · 176

V

valuation
    gap · 19
    level · 48
    targets · 31
value benchmarking · 2
value chain optimization · 138
value creation · 1, 5, 14, 19, 27,
    77, 181
    barriers · 36, 182
    levers · 21, 47, 49
    managers · 43
    obstacles · 2
    office · 190
    potential · 2, 184
    program · 1, 53, 117, 181
    strategies · 1
value gap · 2, 22, 53, 181
value isoquant · 48
value leaders · 13
value-benchmarking · 50
vertical integration · 109
vision · 15
    company · 20, 27
    operationalization · 31

W

weighted average cost of
    capital (WACC) · 151, 197

Z

zero-based cost model · 71

Bibliography

Achleitner, Ann-Kristin, Alexander Bassen: Investor Relations von Wachstumsunternehmen und etablierten Unternehmen im Vergleich (Investor Relations of Growing and Established Companies) in: Knüppel, Hartmut, Christian Lindner (Publishers): Die Aktie als Marke – Wie Unternehmen mit Investoren kommunizieren sollen (How Companies Should Communicate with Investors) 1st Edition, Frankfort 2001.

Andonian, Andre, Ingo Beyer von Morgenstern, Hanno Schmidt-Gothan, et al.: Creating Value in the Network Equipment Industry, McKinsey Telecommunications Autumn 1999 - The Battle for Value. McKinsey & Company, 1999.

Anslinger, Patricia, Steven Klepper and Somu Subramaniam: Breaking Up is Good to Do, in: The McKinsey Quarterly, Volume 1, 1999.

Baetge, Jörg: Bilanzanalyse (Balance Sheet Analysis), Düsseldorf 1998.

Becker, Manfred: Personalentwicklung. Bildung, Förderung und Organisationsentwicklung in Theorie und Praxis (Human Resource Development. Education, Promotion and Organizational Development in Theory and Practical Applications), 2nd Edition, Stuttgart 1999. Blumberg, Donald F.: Managing Service as a Strategic Profit Center, McGraw Hill, New York 1991.

Bogan, Christopher E.: Benchmarking for Best Practices: Winning Through Innovative Adaptation, McGraw Hill, New York 1994.

Brealey, Richard A.: Principles of Corporate Finance, 6th Edition, McGraw Hill, Boston 2000.

Brownell, Robert A.: Planning, Performing, and Controlling Projects: Principles and Applications, 2nd Edition, Upper Saddle River, New York 2000.

Bruner, Robert F.: Case Studies in Finance, Managing For Corporate Value Creation, 3rd Edition, McGraw Hill, New York 1998.

Bühler, Wolfgang: Unternehmenssteuerung und Anreizsysteme (Corporate Control and Incentive Systems) Schäffer-Poeschel, Stuttgart 1999.

Camp, Robert C.: Benchmarking, American Society for Quality, 1995.

von Campenhausen, C., A. Rudolf: Shared Services – profitabel für vernetzte Unternehmen (Shared Services – Yielding Profits for Networked Companies), in: Harvard Business Manager 1/2001.

Chew, Donald H.: The New Corporate Finance, Where Theory Meets Practice, 3rd Edition, McGraw Hill, Boston 2001.

Clausing, Kimberly A.: The Impact of Transfer Pricing on Intrafirm Trade, Cambridge 1998.

Clinton, David; Clancy, Anthony J.; Ziegler, Reinhard; Jensen, Edward W. and Smith, David: The High-Performance Workforce. Separating the Digital Economy's Winners from Losers, Accenture Study 2001.

Clive, Emanuel R.: Transfer Pricing, in: Int. Thomson Business Press, 1994.

Collingwood, Harris: The Earnings Game – Everybody Plays, Nobody Wins, in: Harvard Business Review, June 2001.

Cooper, Robin: The Design of Cost Management Systems, 2nd Edition, Upper Saddle River, New York 1999.

Copeland, Thomas E.: Financial Theory and Corporate Policy, 3rd Edition, Reading 1992.

Copeland, Thomas E.: Valuation, Measuring, and Managing the Value of Companies, 3rd Edition, Wiley, New York 2000.

Crisand, Ekkehard and Pamela Stephan: Personalbeurteilungssysteme. Ziele, Instrumente, Gestaltung (Personnel Evaluation Systems. Objectives, Tools, Design), 2nd Revised Edition, Heidelberg 1999.

Damodaran, Aswath: Corporate Finance, Theory and Practice, 2nd Edition, Wiley, New York 2001.

Dobbs, Richard F.C., Timothy M. Koller: The Expectations Treadmill. In: The McKinsey Quarterly Number 3, 1998.

Eckes, George: Making Six Sigma Last, Managing the Balance Between Cultural and Technical Change, Wiley, New York 2001.

Feinschreiber, Robert: Transfer Pricing Handbook, Wiley, New York 2001.

Fine Charles H.: Clockspeed: Winning Industry Control in the Age of Temporary Advantage, Perseus, 1999.

Fine, Charles. H.: Industry Clockspeed and Competency Chain Design: An Introductory Essay, in: Proceedings of the 1996 Manufacturing and Service Operations Management Conference, Dartmouth College, Hanover 1996.

Fine, Charles: Clockspeed, Wie Unternehmen schnell auf Marktveränderungen reagieren können (Clockspeed, How Corporations Can React Quickly to Market Changes), 1st Edition, Hoffmann & Campe, Hamburg 1999.

Frey, Bruno S.: Markt und Motivation. Wie ökonomische Anreize die (Arbeits-) Moral verdrängen (Market and Motivation. How Financial Incentives Undermine Morals (at Work), 1st Paperback Edition, Munich 1997.

Gerds, Johannes and Schewe, Gerhard: Post Merger Integration: Binsenweisheiten und Erfolgsrezepte (Post Merger Integration: Truisms and Concepts for Success), University of Münster, 2001

Geroski, Paul A.: Innovation, Profitability and Growth over the Business Cycle, London 1992.

Gessner, Karsten: Package-Features für die Kommunikation in den frühen Phasen der Automobilentwicklung. Berichte aus dem Produktionstechnischen Zentrum Berlin. Dissertation. Technische Universität Berlin: 2001. (Package-Features for Communications in the Early Phases of Automotive Development. Reports from the Production Technology Center in Berlin. Dissertation. Technical University Berlin, 2001).

Gleason, Kimberly: The Interrelationship between Culture, Capital Structure, and Performance: Evidence from European Retailers, in: Journal of Business Research, 2000, 50, 185-191

Goldman Sachs Investment Research (Publishers): B2B: 2B or not 2B, Version 1.1, New York 1999.

Greaver, Maurice F.: Strategic Outsourcing, Risk Management, Methods and Benefit, 1999.

Grewe, Alexander: Implementierung neuer Anreizsysteme, Grundlagen, Konzept und Gestaltungsempfehlungen (Implementation of New Incentive Systems. Basics, Concept and Design Recommendations) Hampp, Munich 2000.

Hachmeister, Dirk: Der Discounted Cash Flow als Maß der Unternehmenswertsteigerung (Discounted Cash Flow as a Measure of Company Value Creation), 4th Edition, Frankfurt 2001.

Hausschildt, Jürgen: Innovationsmanagement (Innovation Management), 2nd Edition, Vahlen, Munich 1997.

Hill, Charles W.L.: International Business: Competing in the Global Marketplace, 3rd edition, Boston et al. 2001.

Hungenberg, Harald: Strategisches Management im Unternehmen: Ziele - Prozesse - Verfahren (Corporate Strategic Management: Goals – Processes – Procedures), Gabler, Wiesbaden 2000.

Hunger, J. David: Strategic Management, 7th Edition, Upper Saddle River, New York 2000.

Jetter, Wolfgang: Effiziente Personalauswahl, (Efficient Recruiting) 1st Edition, Stuttgart 1996.

Johnson, Gerry: Exploring Corporate Strategy, 5th Edition, London 1999.

Jones, Gary E., Dirk van Dyke: The Business of Business Valuation, McGraw Hill, New York 1998.

Kaiser, Kevin: Corporate Restructuring & Financial Distress, An International View of Bankruptcy Laws and Implications for Corporations Facing Financial Distress, in: Working Paper, INSEAD, 1994 - quoted in Rajan / Zingales, 1995.

Kaplan, Robert. S.: Using the Balanced Scorecard as a Strategic Management System. In: Harvard Business Review (1996.) 1-2.

Kaplan, Robert S.: Cost & Effect, Using Integrated Cost Systems to Drive Profitability and Performance, Harvard Business School, 1997.

Kaplan, Robert S.: The Strategy Focused Organization: How Balanced Scorecard Companies Thrive in the New Business Environment, Harvard Business Press, 2001.

Kerzner, Harold: Project Management, A Systems Approach to Planning, Scheduling, and Controlling, 7th Edition, Wiley, New York 2000.

Köcher, Wolfgang, Frank Jetter, Ralf Kop and Rainer Skrotzki (Publishers): Managementkompetenz für Führungskräfte, Das Handbuch zur Personalführung und Personalentwicklung (Management Competency for Executives. A Manual for Personnel Management, and Development), LIT, Münster 2000.

Kotler, Philip, Friedhelm Bliemel: Marketing-Management. Analyse, Planung, Umsetzung und Steuerung (Marketing Management. Analysis, Planning, Implementation and Control), 9th Edition, Stuttgart 1999.

Kressler, Herwig W,: Leistungsbeurteilung und Anreizsysteme. Motivation, Vergütung, Incentives (Performance Evaluations and Incentive Systems. Motivation, Compensation, Incentives), 1st Edition, Vienna 2001.

Macharazina, Klaus: Unternehmensführung: das internationale Managementwissen, Konzepte - Methoden – Praxis (Corporate Management: International Management Know-how, Concepts, Methods, Practical Applications), 3rd Edition, Gabler, Wiesbaden 1999.

Mager, Robert F., Peter Pipe: Analyzing Performance Problems, 3rd Edition, Atlanta 1997.

Mei-Pochtler, Antonella: Sharebranding – Die Aktie zwischen objektiver und subjektiver Differenzierung (Sharebranding – Stocks Between Objective and Subjective Differentiation), in: Knüppel, Hartmut and Christian Lindner: Die Aktie als Marke, Wie Unternehmen mit Investoren kommunizieren sollen (The Stock as a Brand: How Companies Should Communicate with Investors), 1st Edition, Frankfort 2001.

Mintzberg, Henry, Bruce Ahlstrand and Joseph Lampel: Strategy Safari, Prentice Hall Europe, 1998.

Myers, Stewart C., Nicholas Majluf: Corporate Financing and Investment Decisions When Firms Have Information That Investors Do Not Have, in: Journal of Financial Economics, 1984, 13, 187-221,

Natusch, Ingo: "Tracking Stock" als Instrument der Beteiligungsfinanzierung diversifizierter Unternehmen (The Tracking Stock as a Funding Tool in diversified Companies), Botemann and Botermann, Cologne 1995.

Pande, Peter S.: The Six Sigma Way, How GE, Motorola, and Other Companies are Honing Their Performance, McGraw Hill, Boston 2000.

Perridon, Louis und Manfred Steiner: Finanzwirtschaft der Unternehmung (Corporate Finance), 9th Edition, Munich 1997.

Porter, Michael E.: Competitive Advantage, Creating and Sustaining Superior Performance, 5th Edition, Free Press, 1999.

Porter, Michael E.: Competitive Strategy, Techniques for Analyzing Industries and Competitors, 10th Edition, Free Press, 1998.

Rajan, Raghuram and Luigi Zingales: What Do We Know About Capital Structure? Some Evidence from International Data, Journal of Finance, 1995, Vol L, No 5, pp1421ff

Rappaport, Alfred and Mark Sirower: Unternehmenskauf – mit Aktien oder in bar bezahlen? (Corporate Acquisitions – Pay in Cash or in Stocks?) In: Harvard Business Manager, March 2000.

Schuler, Andreas and Pfeifer, Andreas: Efficient eReporting with SAP EC, Strategic Direction and Implementation Guidelines, Vieweg Verlagsgesellschaft, 2001

Smith, Ian G.: Incentive Schemes, People and Profits, Croner, London 1990.

Stadtler, Hartmut und Christoph Kilger: Supply Chain Management and Advanced Planning, Concepts, Models, Software and Case Studies, 2000.

Stewart, G. Bennet und Al Ehrbar: EVA, The Real Key to Creating Wealth, John Wiley & Sons, New York 1998.

Tang, Roger Y. W.: Intra-firm Trade and Global Transfer Pricing Regulations, Quorum Books, 1997.

Weber, Jürgen: Balanced Scorecard & Controlling: Implementierung - Nutzen für Manager und Controller - Erfahrungen in deutschen Unternehmen (Balanced Scorecard & Controlling: Implementation - Benefits for Managers and Controllers – Experiences in German Companies), 3rd Edition, Gabler, Wiesbaden 2000.

Welge, Martin K.: Strategisches Management: Grundlagen - Prozess – Implementierung (Strategic Management: Basics – Process – Implementation), Gabler 1999.

West, Thomas L.: Handbook of Business Valuation, 2nd Edition, Wiley, New York 1999.

White, Michelle J., Corporate Bankruptcy: A U.S.-European Comparison, in: Working Paper, University of Michigan, quoted in Rajan / Zingales, 1995

Wideman, R. Max: Project & Program Risk Management, A guide to Managing Project Risks and Opportunities, Project Management IstPubns, 1998.

Wiedemann, Anja and Michael Paschen: Personalentwicklung, Potentiale ausbauen, Erfolge steigern, Ergebnisse messen (Human Resource Development, Expanding Potentials, Increasing Success, Measuring Results), 1st Edition, Freiburg 2001.

Williams, Michael: War for Talent, 1st Edition, London 2000.

Consulting Books

Nikolaus Mohr/Gerhard P. Thomas
Interactive Broadband Media
A Guide for a Successful Take-Off
2001. xii, 177 pp. with 48 figs. Hardc. € 74,00   ISBN 3-528-05781-5
„Das Buch bietet eine fundierte Analyse des Markts der Breitband-
medien.“   Kress-Report 38/01

Andreas H. Schuler/Andreas Pfeifer
Efficient eReporting with SAP EC®
Strategic Direction and Implementation Guidelines
2001. x, 217 pp. with 112 figs. (Efficient Business Computing, ed. by
Fedtke, Stephen) Hardc. € 99,00   ISBN 3-528-05761-0
Contents: Management reporting and legal consolidation Intgration of
management reporting and legal consolidation and integration roadmap
- Reporting and the role of information technology - Efficient Reporting
with SAP R/3 ES - Design, Implementation, Test, and Deploy -
eReporting

Heinz-Dieter Knoell/Lukas W. H. Kuehl/ Roland W. A. Kuehl/
Robert Moreton
Optimising Business Performance
with Standard Software Systems
How to reorganise Workflows by Chance of Implementing new
ERP-Systems (SAP, BAAN, Peoplesoft, Navision, ...) or new Releases
2001. xvi, 425 pp. with 191 figs. Hardc. € 74,00   ISBN 3-528-05765-3

vieweg

Abraham-Lincoln-Straße 46
65189 Wiesbaden
Fax 0611.7878-400   Stand 15.3.2002. Änderungen vorbehalten.
www.vieweg.de   Erhältlich im Buchhandel oder im Verlag.

4